Advanced Manufacturing

Springer
London
Berlin
Heidelberg
New York
Barcelona
Hong Kong
Milan
Paris
Singapore
Tokyo

Other titles published in this series:

Modern Manufacturing
M.B. Zaremba and B. Prasad (Eds)

Advanced Fixture Design for FMS
A.Y.C. Nee, K. Whybrew and A. Senthil kumar

Intelligent Quality Systems
D.T. Pham and E. Oztemel

Computer-Assisted Management and Control of Manufacturing Systems
S.G. Tzafestas (Ed.)

The Organisation of Integrated Product Development
V. Paashuis

Advances in Manufacturing:: Decision, Control and Information Technology
S.G. Tzafestas (Ed.)

Computer Applications in Near Net-Shape Operations
A.Y,C. Nee, S.K. Ong and Y.G. Wang (Eds)

Parallel Kinematic Machines
C.R. Böer, L. Molinari-Tosatti and K.S. Smith (Eds)

Manufacturing and Supply Systems Management
B. Wu

Robot Manipulation of Deformable Objects
D. Henrich and H. Wörn (Eds)

Hierarchical Operations in Supply Chain Planning
Tan Miller

E. Stanley Lee and Hsu-shih Shih

Fuzzy and Multi-level Decision Making

An Interactive Computational Approach

With 8 Figures

Springer

Professor E. Stanley Lee
Department of Industrial & Manufacturing Systems Engineering, Kansas State University, Manhattan, Kansas 66506, USA

Assistant Professor Hsu-shih Shih
Graduate School of Management Science, I-Shou University, Ta-Hsiang, Kaohsiung 84008, Taiwan

Series Editor

Professor Duc Truong Pham, PhD, DEng, CEng, FIEE
University of Wales Cardiff
School of Engineering, Systems Division,
P.O. Box 917, Cardiff CF2 1XH, UK

ISBN-13:978-1-4471-1177-1

British Library Cataloguing in Publication Data
Stanley, Lee E.
Fuzzy and multi-level decision making : An interactive computational approach. - (Advanced manufacturing series)
1.Programming (Mathematics) 2.Fuzzy logic
I.Title II.Hsu-Shih, Shih
519.7'03
ISBN-13:978-1-4471-1177-1

Library of Congress Cataloging-in-Publication Data
Lee, E. Stanley (Eugene Stanley), 1930-
Fuzzy and multi-level decision making : an interactive computational approach / E. Stanley Lee and Hsu-Shih Shih.
p. cm. -- (Advanced manufacturing series)
Includes bibliographical references.
ISBN-13:978-1-4471-1177-1 e-ISBN-13:978-1-4471-0683-8
DOI: 10.1007/978-1-4471-0683-8

1. Soft computing. 2. Mathematical optimization. 3. Operations research. I. Shih, Hsu-Shih, 1955- II. Title. III. Series.
QA76.9.S63L44 1999
006.3--dc21 99-21135

Softcover reprint of the hardcover 1st edition 2001

Typesetting: Camera ready by authors

69/3830-543210 Printed on acid-free paper SPIN 10713346

Preface

Two types of decision-making process usually encountered in a hierarchy organization are discussed in this book. The duo-ploy or multi-ploy game-like decision making process, also known as bi-level or multi-level programming, and the fuzzy interactive decision-making approach. The former is a well-known approach and has been applied to many practical duo-ploy and multi-ploy problems, especially in economic systems. The latter is proposed to overcome the computational difficulties by exploring the inherent fuzzy nature of a large organization. Much more research is needed, both from the theoretical and computational standpoints, for the fuzzy interactive approach.

The basic concept of multi-level decision making is that an upper-level decision-maker sets his or her goal and/or decision and then asks each subordinate level of the organization for their optima. The decisions of the lower levels are then submitted and modified by the upper-level with consideration of the overall benefits of the organization. This mutually interactive process is continued until reaching a solution, which is satisfactory to all the decision-makers. Obviously, the degree of interaction and the degree of satisfaction depend on the management style of the upper level. This decision-making process is extremely useful to the hierarchy decentralized organizations such as the various manufacturing and service companies.

Although there exist many optimization tools such as the decomposition principle, goal programming, multi-objective programming, and game theory, almost all of these traditional approaches cannot meet the common features of the decision process of a multi-level decentralized organization. Duo-ploy or multi-ploy game-like decision making process, also known as bi-level or multi-level programming, is developed for hierarchy systems. However, these approaches generally assume that the lower level decision is final and no further interaction is required. Although we could modify the procedure to allow continuous interactions, the computational requirements would be tremendous if continuous interaction was carried out.

In order to increase the computational efficiency of the basic multi-level programming algorithms, a completely different philosophy of exploring the typical fuzziness, vagueness, or the not-well-defined nature of a large decentralized hierarchy organization using fuzzy set theory was proposed in this research. Although much more research is needed, the resulting fuzzy interactive sequential approach appears to be a useful and powerful one. The advantages are that not only are the computational requirements are reduced tremendously, the representation of the system is also more realistic. In other words, the traditional approach is trying to solve a non-existing problem by requiring an unrealistically accurate model and by ignoring the inherent fuzziness of large organizations.

The book can be divided approximately into three parts. The first part, which includes Chapters 1, 2, and 3, summarizes the multi-level programming algorithms. Obviously, it is impossible to cover all the approaches, only the better known ones are summarized. The emphasis is on the numerical solution aspects and no theoretical treatment is included.

The second part, which includes Chapters 4 and 5, summarizes knowledge representation and fuzzy decision making. Because of the frequent use of linguistic expressions in the interactions between the various levels of management in a hierarchy organization, emphasis is placed on linguistic representation by the use of fuzzy concept. In here, we are indebted to Professor Zadeh and most of the presentations are based on Professor Zadeh's writings.

The third part, which includes the chapters after Chapter 5, presents the interactive decision-making algorithm. Chapter 6 studies the basic interactive algorithms, Chapter 7 considers the different aspects of aggregation, and Chapter 8 illustrates a practical application to solve the minimum cost-flow problem.

Several examples and algorithms are adopted from the original publications as acknowledged in the text. In particular, we are grateful for the following permissions from the copyright owners: INFORMS, Table 2.4; Baltzer Science Publishers, Figure 2.3; Society for Industrial and Applied Mathematics, Figure 2.4; IEEE, Table 3.1; and Elsevier, Figure 4.2.

One of the author wishes to express his appreciation to the colleagues in Industrial and Manufacturing Systems Engineering of Kansas State University, and, especially to the Department Chairman, Professor Bradley A. Kramer, for providing the environment to accomplish this work. We also wish to express our appreciation to the editorial staff of Springer-Verlag London Ltd., and especially to Hannah Ransley, for the tremendous amount of help in overcoming the software problem.

Table of Contents

Chapter 1

Introduction

In this chapter, two important types of hierarchical decision-making processes will be introduced by considering the actual decision-making in hierarchy systems. The two types, which form the basic topics of this book, are the multi-ploy game-like decision-making process and the interactive decision-making process. The former, also known as multi-level programming, is important because of the presence of many practical problems, which are, or can be reformed into, the multi-ploy type. This is especially true in the duo-ploy economic systems. The interactive decision making is generally applicable in large hierarchy organizations, which are characterized by the mutual interactions in a top-to-bottom sequence and by the vague and not-well-defined nature of a large hierarchy organization. Since there are many algorithms for solving the multi-ploy type decision making process, an approximate classification is proposed. Furthermore, for the use in latter chapters, where some of the algorithms will be presented, the basic characteristics such as the Nash equilibrium and the Stackelberg-Nash equilibrium of the duo-ploy systems are also discussed.

1.1 Decision Making in Hierarchy Systems: Multi-ploy versus Interactive Decisions

For simplicity, let us first restrict our discussion to two-level hierarchy system. Furthermore, the decision-maker (DM) of the top or first level will be designated as the leader and the DM of the second or lower level, the follower. Depending on the degree of interaction or cooperation between the two levels, various decision-making processes can be formulated. At one extreme, the objective of the follower is in direct opposition to the objective of the leader and thus the problem is reduced to the classical max-min problem. At the other extreme, the follower is in complete cooperation with the leader. However, most problems encountered in practice falls between these two extremes. The principal aim of this research is concerned with these practical problems in between, among which, the two most important processes are: (1) a decision process which is somewhat similar to the static two person Stackelberg game and (2) the interactive decision-making process. For the convenience of discussion, the first type of decision process will be referred to as the duo-ploy decision-making process. The corresponding decision making process for multi-level hierarchy system will be referred to as multi-ploy decision making. Besides the practical reasons mentioned above, the separation of the multi-ploy process is also due to historical developments. In fact, many algorithms, which are known as the bi-level programming problem (BLPP) algorithms, have been developed for solving these duo-ploy processes.

To illustrate the characteristics of the multi-ploy process, let us consider a process with two levels, or the duo-ploy system. The decision of a typical duo-ploy process is carried out in the following manner. The upper-level DM, the leader, makes his or her decision first with full information about both levels, and then the second level DM, the follower, makes his or her decision in isolation and based on the first level's decision. In most practical cases, the second-level decision is the final decision and no time or energy for the upper level to make another decision afterward. Notice that although the second level cannot control the decision of the first level, the final decision of the lower does eventually influence the upper level and the overall results. There are many practical examples of this game type of interactions. For example, consider the traffic-planning problem [Migdalas 1995]. The planner or the decision-maker, the leader, seeks to improve the performance of the traffic network, while at another level the network users, the followers, make choice with regard the details of their travel based on the network formed by the planner. Another example is the taxation problem, where the government, the leader, decides the taxation rules and rates, while the taxpayers, the followers, try to minimize their payments based on these rules or rates. A third example is the pricing and purchasing of fertilizers within an agricultural region [Fortuny-Amat and McCarl 1981]. As a final example, even the issue of setting penalties for illegal drug importation can be formulated in terms of the duo-ploy problem [Candler and Townsley 1982]. Three- or multi-ploy problems, although much more difficult to solve, also occur frequently in practice. One example is the allocation of tax money by the federal government to the various states, which in turn allocate to different cities, and which, again, in turn allocate to different project managers. Notice that, in all the above examples, the leader has to consider both his immediate gains and the influences of his decision on the followers, whose final decision would ultimately influence the gains of the leader and the overall system. Under these situations, the follower's decisions are usually final. In actual practice, the planner or the policy maker cannot change their decision after examining the followers' decision. It is true that the network or the policy could be changed. However, in practice these changes seldom immediately carried out.

Besides the above examples, multi-ploy type structure can be found in diverse areas such as economic systems, ecology and environmental studies, biology and chemical engineering, network design, transportation, game theory, data bases, and the theory of classification.

Obviously, the multi-ploy strategy in decision making can seldom be completely adapted without change by a hierarchy organization such as a manufacturing or a service company. In general in a multi-hierarchy company, after the followers or the divisions of the company made their decisions based on the leader's decision, the divisions must submit their decisions and then modified by the upper-level DM with considerations of the overall benefits for the organization. Ideally, this mutually interactive decision process is continued until a satisfactory solution or compromise is reached. This continuously interactive process will be referred to as the interactive decision making process, which is listed as the second process previously. Obviously, depending on the management style, the degree of cooperation can be different for different companies. This kind of decision making process is extremely practical and can be applied to the various decentralized large hierarchic companies. Unfortunately, this practical decision making approach cannot be solved easily even for the bi-level organization. One of the aims of this book is to explore the recently developed interactive fuzzy approach to solve this multi-level interactive decision problem.

Although the decision making process in a hierarchy organization, such as in the service or manufacturing industries, is usually not completely like the multi-ploy decision process, it is also seldom completely controlled by the top management. In general, the top management is more concerned with the overall strategy, the market share and compatibility and not with the detailed day-to-day activities. Thus, it rarely exercises direct control over its divisions, although its general policy will influence and partially control the decision processes of the divisions. The division managers can make independent decisions as long as not in conflict with those of the top management. Because of this absence of a central decision-maker, different degrees of interactions between the various hierarchy levels can be formed with the aim to increase the overall organization profit.

1.2 Bi-level and Multi-level Programming

As mentioned before, the traditional approach to solve the duo-ploy and multi-ploy decision problems are known as the bi-level and multi-level programming problems, respectively. Bracken and McGill [1973] seem to be the first researchers to investigate and to define this hierarchy decision problem as a generalized mathematical programming problem. Since then, various investigators have studied this multiple-level programming problem. In order to summarize the results, it is convenient to classify the problem based on both the structural complexity and the approach of the solution method. Structurally, this programming problem can be classified into bi-level programming problem (BLPP), bi-level decentralized programming problem (BLDPP), multi-level programming problem (MLPP), and multi-level decentralized programming problem (MLDPP). A decentralized problem has multiple divisions in each level except the highest level and a BLPP or a MLPP has only one division in each level. Of course, MLPP can be further subdivided into three-level programming problem (TLPP), four-level programming problem, and *etc*. Based on the solution approach, the solution methods can be roughly classified into four categories as shown in Table 1.1. The first three categories started relatively early and are based on traditional optimization or decision concepts and the last one is based on more recent developments in decision-making and thus more decision-making than optimization. For the

Table 1.1. Bi-level and multi-level programming

Classification	Methods	References
I. Extreme-point search	k^{th}-best algorithm	Bialas and Karwan [1982, 1984] Wen and Bialas [1986]
	Grid-search algorithm	Bard [1982, 1983b]
II. Trans-formation approach	Complementary pivot	Judice and Faustino [1988. 1992] Bialas and Karwan [1984] Candler and Townsley [1982] Bialas *et al.* [1980], Onal [1993]
	Branch-and-bound	Bard and Falk [1982] Bard and Moore [1990] Hansen *et al.* [1992] Fortuny-Amat and McCarl [1981]
	Penalty function	Anandalingam and White [1990, 1993], Aiyoshi and Shimizus [1981a, 1981b, 1984] Anandalingam and Apprey [1991]
III. Descent and heuristic	Descent method	Savard and Gauvin [1994]
	Branch-and-bound	Bard and Moore [1990]
	Cutting plane	Bard [1984b]
IV. Evolutionary approach	Tabu search	Grendreau *et al.* [1996] Wen and Huang [1996]
	Genetic algorithm	Liu [1998] Mathieu *et al.* [1994]

convenience of discussion, the first three categories will be referred to as the classic or traditional approaches. The first two categories, namely extreme-point search and transformation approaches, are developed solely for solving linear or linear-quadratic bi-level programming problems based on traditional optimization concepts. Some of them were also extended to linear BLDPP and TLPP. The basic concept of extreme-point search is to seek a compromise vertex by simplex algorithm based on adjusting the control variables. The transformation approach involves transforming the lower-level problems into constraints for the higher level by the use of various techniques such as the Karush-Kuhn-Tucker (KKT) conditions, penalty functions, *etc.* This transformed problem becomes the one-level mathematical programming problem. Because of the appearance of non-linear terms, the reformulated problem in the transformation approach becomes complex, *non-linear* and difficult to solve numerically. However, several fairly efficient numerical iterative techniques have been proposed.

Although Categories I and II can be extended or modified for *non-linear* problems, Categories III is developed solely for discrete or *non-linear* BLPPs and is based on existing search or heuristic decision making approaches such as gradient technique, cutting-plane algorithm, and branch-and-bound heuristics. In general, the problem is non-convex, thus optimality cannot be guaranteed even when the search procedure appears to be exhausted.

Category IV is based on the recently developed evolutionary approaches, which is especially suited for solving NP-hard problems such as the multi-ploy problems. This application is relatively new and has very great potential. Tabu search [Glover 1990] has been used to solve both the linear BLPP[Gendreau *et al.* 1996] and the mixed- integer linear BLPP [Wen and Huang, 1996]. Mathieu *et al.* [1994] proposed the use of simulated annealing to solve this NP-hard problem. Liu [1998] illustrated the use of genetic algorithm to solve the Stackelberg-Nash equilibrium for multi-level programming problems.

Some typical references for each category are also listed in Table 1.1. This list is certainly not complete. Most of the computational algorithms are developed for the simplest linear BLPPs. But, unfortunately, even for this simplest problem, there exist no simple solutions. In fact, it has been proved that even for this simplest two-level linear case the problem is strongly *NP*-hard. Furthermore, the geometric properties of the bi-level linear problem are much more complicated than the usual mathematical programming problem and thus the set of feasible solution is non-convex and non-unique. Another difference is that no general hypothesis on cost function, which will guarantee Pareto optimal, can be obtained even for the linear BLPPs unless both the upper and lower objectives coincide, in which case, both DMs completely cooperate and lead to the optimal solution of BLPP which is Pareto optimal. Thus, the solutions of BLPP or MLPP are, in general, not Pareto optimum.

The difficulties in obtaining an effective solution for the multi-ploy problems are two folds: the complexity caused by the interactions of the various DMs in the various levels and the implicit nature caused by these interactions. Due to the imbedding of the top-level problems into the lower levels, most optimization or decision making techniques are too restrictive and cannot meet the flexibility demands. Furthermore, because of the limited freedom of the lower decision-makers, there easily exists no solution of the overall problem.

1.3 Characteristics of Duo-ploy Systems

To investigate the basic characteristics of a duo-ploy game type or leader-follower decision process, consider the following bi-level problem:

$$\max_{x1} f_1(x_1, x_2) = c_{11}^T x_1 + c_{12}^T x_2 \qquad \text{(upper level)} \quad (1.1)$$

where x_2 solves:

$$\max_{x2} f_2(x_1, x_2) = c_{21}^T x_1 + c_{22}^T x_2 \qquad \text{(lower level)} \quad (1.2)$$

subject to:

$$A_1 x_1 + A_2 x_2 \leq b$$
$$x_1 \geq 0$$
$$x_2 \geq 0$$

where A_1 and A_2 are $m \times n_1$- and $m \times n_2$-dimensional matrices, respectively; c_{11}, c_{21} and x_1 are n_1-dimensional vectors; c_{12}, c_{22} and x_2 are n_2-dimensional vectors; and b is an m-dimensional vector; f_1 and f_2 are objective functions to be optimized by the leader and the follower, respectively; and x_1 and x_2 are decision variables controlled by the leader and the follower, respectively.

The decision making process is carried out in a leader to follower sequence. The leader makes the decision first and then the follower must make the decision based on the leader's decision, or based on x_1, which has already decided by the leader. The game is static in the sense that the follower's decision is final and the leader cannot make another decision after the follower's decision. The problem can be interpreted in another way. Since x_2 is really a function of x_1, the problem can be formulated as a parametric programming problem, in which the follower must solve a problem parameterized in the leader. To examine the basic characteristics of the problem, we need to introduce the following generally used definitions:

Follower's feasible region with fixed $x_1 \in X_1$:

$$\Omega(x_1) = \{x_2 \mid x_2 > 0, A_1 x_1 + A_2 x_2 \leq b\} \quad (1.3)$$

Follower's rational reaction set:

$$\Psi(x_1) = \{x_2 \mid x_2 \text{ solves: } \max_{x2} [f_2(x_1, x_2) \text{ with } x_2 \in \Omega(x_1)]\} \quad (1.4)$$

Inducible region:

$$\text{IR} = \{(x_1, x_2) \mid x_1 \in X_1, x_2 \in \Psi(x_1)\} \quad (1.5)$$

Notice that follower's rational reaction set is in response to a fixed x_1 and it is the best decision set for the follower only. The inducible region is different from the feasible region of the BLPP in that the IR must be within the optimal set of the follower. Of course, the IR is a sub-set of the feasible set of the problem. Furthermore, the IR represents the set, over which the leader may optimize if the leader has the control over all the variables.

Because of the leader-follower decision sequence, the problem is much more complicated. For example, even if we assume (1) the feasible region of the BLPP is non-empty and bounded, and (2) $\Omega(x_1) \neq 0$, that is, for each decision of the leader, the follower has a feasible response; the problem may still have difficulty. This is principally caused by the fact that the leader cannot response after the follower's decision. If the follower's rational reaction set is not a singleton, then different values of x_2 may produce different values of the leader's objective function. It is most probable that not all the different values of x_2 will maximize f_1. Most of the algorithms using the Karush-Kuhn-Tucker condition to imbed the lower problem into the leader's problem and thus exploiting the set of the inducible region. However, if the algorithm strictly follows the decision sequence or the algorithm cannot take care of the case where the follower's rational reaction set is not a singleton, the above problem of multiple solutions of the lower decision maker must be considered with care.

1.4 Characteristics of Duo-ploy Systems with Multi-followers

For duo-ploy system with multiple followers, which is also known as the BLDPP, the problem is even more complicated. Consider the *non-linear* BLDPP with p followers:

$$\max_x F(x, y_1, y_2, \ldots, y_p) \tag{1.6}$$

subject to:

$$G(x) \leq 0$$

where each y_i, $i = 1, 2, \ldots, p$, solves:

$$\max_{xi} f_i(\mathrm{x}, y_1, y_2, \ldots, y_\mathrm{p}) \tag{1.7}$$

subject to:

$$g(x, y_1, y_2, \ldots, y_p) \leq 0$$

where x is the control of the leader and $y_1, y_2, \ldots, y_p$ are the controls of the p followers, and f_i is the objective function of the ith follower.

We shall assume that the followers know the leader's strategy and also the followers reveal their strategies simultaneously to each other. After the leader and other followers reveal their strategies, the reaction of the ith follower's optimal solution, y_i^*, is obtained by solving:

$$\max_{yi} f_i(x\, y_1, y_2, \ldots, y_p),\ i = 1, 2, \ldots, p \tag{1.8}$$

subject to:

$$g(x, y_1, y_2, \ldots, y_p) \leq 0$$

Equation 1.18 may have multiple solutions. Following the definition used in the previous section, these solutions are known as the rational reaction set of the ith follower. Since all the followers have equal status and each follower optimizes his own objective individually; the well-known Nash equilibrium solution, which obeys the following optimality condition, is obtained:

$$f_i(x, y_1^*, \ldots, y_{i-1}^*, y_i, y_{i+1}^*, \ldots, y_p^*) \leq f_i(x, y_1^*, \ldots, y_{i-1}^*, y_i^*, y_{i+1}^*, \ldots, y_p^*) \qquad (1.9)$$

for $i = 1, 2, \ldots, p$. The array $(y_1^*, \ldots, y_{i-1}^*, y_i^*, y_{i+1}^*, \ldots, y_p^*)$ is known as the Nash equilibrium with respect to x. In general, Nash equilibrium is not unique and is also not efficient in the sense of Pareto optimum. Thus, anyone of the followers can move to a better position by deviating from the Nash equilibrium.

In Nash equilibrium, only the followers obtained their best solutions with respect to any given $x \in X$. The Stackelberg-Nash equilibrium is the array $(x^*, y_1^*, y_2^*, \ldots, y_p^*)$, which also contains the leader's optimum. In other words, the Stackelberg-Nash equilibrium is essentially the optimal solution of the BLDPP problem.

1.5 Fuzzy Interactive Decision Making

As has been discussed in the previous section, most of the algorithms proposed during the last two decades are for solving the simplest bi-level problems of the duo-ploy type and none of them are computationally efficient for large hierarchy organizations. In fact, they are not very efficient even for bi-level problems. Recently, a completely different approach by exploring the typical fuzziness, vagueness, or the not-well-defined nature of a large hierarchical organization using fuzzy set theory was proposed by Lee and coworkers [Shih *et al.* 1996 and Shih and Lee 1999]. The resulting fuzzy interactive sequential approach has been proved to be a powerful one and can be used to help the decision-maker to solve practical problems encountered in large decentralized companies. The advantages of the proposed approach are two folds: First, the problem becomes much more simplified and thus it can be solved reasonably easily for fairly large practical problems and, secondly, the representation of the original problem is not only simplified but also much more realistic. In other words, since the real world problems for large organizations are generally fuzzy or not well defined, the existing classic algorithms are trying to solve a non-existing problem by assuming unrealistically accurate models and by ignoring the inherent fuzziness of large organizations.

The fuzzy interactive approach is formulated by the combined use of the concept of fuzzy tolerance membership functions and multi-objective decision making. The basic idea is to model an approximate system approximately and followed by iterative improvements or supervised search to obtain a better model. The higher-level DM defines his or her objectives and decisions first with possible tolerances represented by fuzzy membership functions. This tolerance information of the upper-level constraints the lower-level DM's feasible space. The lower-level DMs also form individual fuzzy membership functions based on their tolerance levels. A search is carried out based on sequentially improving the membership functions, which represent the satisfactory degrees of both the upper- and lower-level decision-makers. In order to satisfy all the DMs, multiple-objective programming or other interactive approaches are used to obtain satisfactory membership functions. Unlike the extreme-point search procedure, we do not presume the optimal solution is at a corner point. Since true optimum cannot be defined easily due to the interactions, there is no reason to assume the optimum is at the corner point. In fact, for multiple-objective optimization, a non-dominated solution is the desired one. Thus, we consider a compromise solution, which is acceptable to all the DMs, is acceptable. The proposed approach is very efficient and does not increase the complexity or the size of the original problem. Since the proposed procedure is very robust, it can be extended easily to non-linear parameter multi-level problems with convex property.

In addition to the interacting aspects of the DMs of a large decentralized hierarchical organization, another reason for the vagueness or not-well-defined nature is the frequent use of linguistic terms. For example, it is common for the DM to describe the current earnings or the demand of a certain product as "extremely good, good, average, bad, or extremely bad". Or, give directions such as "put more emphasis on the environment and worker's safety". The fuzzy set theory is ideally suited for representation and for manipulation of these linguistic modifiers.

Chapter 2

Linear Bi-level Programming

As has been pointed out in Chapter 1, most of the developments on multiple-level programming problem (MLPP) start with this structurally simplest bi-level programming problem (BLPP). Another reason is that many decision problems encountered in practice can be formulated as the duo-ploy type decision problem. Thus, various algorithms have been developed for solving these practical problems. Obviously, it is impossible to cover all the algorithms. Only several typical algorithms to illustrate the approaches will be discussed in this chapter. We shall restrict our discussions to the various numerical solution algorithms and avoid any theoretical treatments. Only linear and non-fuzzy approaches or Categories I and II in Table 1.1 will be discussed in this chapter and fuzzy interactive approaches, which form the main aim of this book, will be discussed in detail in later chapters. There are no linear BLPPs in Category III and Category IV does not belong to the traditional optimization approaches. Non-linear and other MLPPs will be discussed in Chapter 3.

2.1 Linear Bi-level Programming

BLPP is usually viewed as a problem with two DMs at two different hierarchical levels. The upper-level decision maker, the leader, selects his or her decision vector first and the lower decision maker, the follower, select his or her afterward based on the decisions of the upper level. The leader knows the functions of the follower, who may or may not know the functions of the leader. This duo-ploy type bi-level problem has been studied extensively in the field of economics, where social objectives versus the objectives of the individual economic agents form a typical Stackelberg's leader-follower duo-ploy model. Thus, the bi-level linear programming problem is often viewed as a special case of two-person, non-zero sum non-cooperative game in which one DM, the leader, has the ability to enforce his/her strategy on the follower. The problem forms a nested optimization model involving two problems, an upper one and a lower one. In a way, the upper-level decision is imbedded in the lower-level decision. The main problem is the aggregation of these imbedded two problems.

To formulate the BLPP, let us introduce the decision vectors, $\boldsymbol{x}_1$ and $\boldsymbol{x}_2$, where the top-level DM has control over the vector $\boldsymbol{x}_1$ and the lower-level DM has control over the vector $\boldsymbol{x}_2$. We shall assume that the performance functions, f_1 and f_2, for the two DMs are linear and bounded. The BLPP can be stated as:

$$\max_{x1} f_1(x_1, x_2) = c_{11}^{\mathrm{T}} x_1 + c_{12}^{\mathrm{T}} x_2 \qquad \text{(upper level)} \quad (2.1)$$

where x_2 solves:

$$\max_{x2} f_2(x_1, x_2) = c_{21}^{\mathrm{T}} x_1 + c_{22}^{\mathrm{T}} x_2 \qquad \text{(lower level)} \quad (2.2)$$

subject to:

$$(x_1, x_2) \in X = \{(x_1, x_2) \mid A_1 x_1 + A_2 x_2 \leq b, \text{ and } x_1, x_2 \geq 0\}$$

where A_1 and A_2 are $\mathrm{m} \times \mathrm{n}_1$- and $\mathrm{m} \times \mathrm{n}_2$-dimensional matrices, respectively, c_{11} and c_{21} are n_1-dimensional vectors, c_{12} and c_{22} are n_2-dimensional vectors, and b is an m-dimensional vector.

It should be noted that, in general, x_2 is a function of x_1, or $x_2(x_1)$. This is because of the fact that x_2 is chosen after those of x_1. Various degrees of cooperation between the leader and the follower can be assumed. From the point of view of the leader, the best case is when the follower accepts the leader's preferences, the cooperative case, and the worst case is the follower adopts the opposite of the leader's preferences.

Notice also that the term $c_{21}^{\mathrm{T}} x_1$ in the objective function of the lower DM can be omitted without influencing the optimum. This is because the decision of the lower DM must based on the already decided x_1.

Because of the various applications in different fields, different names have been used in the literature for the leader. Some of them are "upper", "outer", "level one", or "policy". Similarly, "lower" "inner", "level two", or "behavioral" are used instead of "follower".

Notice that once x_1 is decided, the lower DM's problem becomes a regular mathematical programming problem. Thus, for a given x_1, let $G(x)$ denote the set of optimal solutions of the following lower-level problem:

$$\max_{x2 \in f1(x1)} f_2(x_2) = c_{22}^{\mathrm{T}} x_2 \qquad (2.3)$$

where $f_1(x_1) = \{x_2 \mid A_2 x_2 \leq b - A_1 x_1, x_2 \geq 0\}$ represents the lower-level DM's feasible decision space.

The set of rational reaction for the lower DM or f_2 over the decision x_1 was given in Chapter 1. This set of rational reactions can also be expressed as:

$$\Psi(x_1) = \{x_2 \mid x_2 \in G(x_1)\} \qquad (2.4)$$

Multiple local optima may exist for this bi-level problem and on the other hand, there is the possibility of no solution, which is especially true if both upper and lower problems have constraints. Computationally, the basic problem is the maximization of a piece-wise linear function over a polyhedron, which is formed by the constraints. However, due to the interactions of the two levels, this linear optimization problem cannot be solved easily. Thus, various transformation approaches, which transform the two-level into a single-level problem, have been proposed. Recently, the various evolutionary or local search approaches have also been proposed to solve this very NP-hard problem. Thus, the various algorithms developed for solving the linear BLPPs can be classified into the following four categories:

1. extreme-point search method, which searches over the polyhedron formed by the constraints;
2. transformation method, which transforms the problem into a one-level optimization problem by the use of the Karush-Kuhn-Tucker (KKT) theory or other methods such as penalty function;
3. the evolutionary methods;
4. the fuzzy interactive approach.

Combinations of these various approaches have also been proposed. Table 1.1 listed some of the references for the various classifications. The first two approaches will be discussed in this chapter.

The extreme-point-search approach seeks a compromise vertex along the polyhedron based on the simplex algorithm. It is very inefficient, especially for large problems. To improve the algorithm, short-cut [Bard 1983b] has been proposed. However, with these short cuts, the generalities of the original algorithm appear to be lost.

The transformation approach transforms the original two problems into one by treating the lower-level problem as constraints for the higher level and by using the KKT optimality conditions or other transformation functions. The resulting problem becomes non-linear and complex. However, several fairly effective algorithms have been proposed to solve this problem.

The first step in the transformation approach is to use the KKT conditions to transform the original problem to a first-level auxiliary problem. The resulting problem is equivalent to the elimination of Equation 2.2. Thus, the BLPP can be re-formulated as the following first-level auxiliary problem:

$$\max_{x1,x2} f_1(x_1, x_2) = c_{11}^{\mathrm{T}} x_1 + c_{12}^{\mathrm{T}} x_2 \tag{2.5}$$

subject to:

$$\begin{aligned} &A_1 x_1 + A_2 x_2 \le b \\ &w^{\mathrm{T}} (A_1 x_1 + A_2 x_2 - b) = 0 \\ &w^{\mathrm{T}} A_2 = c_{22} \\ &x_1, x_2, w \ge 0 \end{aligned}$$

which was obtained by applying the KKT optimality conditions with w as the dual row vector or the Lagrange multiplier vector.

For simplicity, we shall assume that the linear BLPP has a solution, which means that the feasible region of the problem with the given constraint is nonempty and, for each decision taken by the upper level, the second level has space to respond within his or her feasible region. However, as has been discussed in the previous chapter, even with these assumptions, the BLPP of the duo-ploy type may still have complications. One example is when the lower-level rational reaction set is not single valued, in which case, depending on the reaction of the lower DM, the upper-level DM may not obtain the best value for his or her objective function.

Due to the complementary slackness condition in the constraints, the transformed problem forms a non-convex non-linear programming problem. The various algorithms proposed to solve this non-linear programming problem can be approximately divided into the following groups:

1. complementary-pivot algorithm. First proposed by Bialas, Karwan and Shaw [1980] and Bialas and Karwan [1984]. Further developments by Judice and Faustino [1988, 1992], Onal [1993], Candler and Townsley [1982]], and others;
2. branch-and-bound algorithm. Investigators who use this approach are, Bard and Falk [1982], Bard and Moore [1990], Fortuny-Amat and McCarl [1981], Hansen, Jaumard and Savard [1992], and others;
3. penalty-function approach. Various schemes have been proposed to define the penalty function. For example, the penalty function can be incorporated with the lower-level objective function only. Or, both objective functions can be penalized. Another example is due to White and Anandalingam [1993]. These authors used a duality-gap-penalty-function format to solve the bilinear version of the linear BLPP.

The complementary-pivot algorithm solves the system by iteration. The basic procedure can be viewed as an implicit enumeration of the lower level, and thus, the solution procedure can be extended easily to the three-level programming problem (TLPP), which will be discussed in the next chapter. However, Ben-Ayed and Blair [1990] showed that the early complementary-pivot algorithm such as that due to Bialas and Karwan [1984] is not guaranteed to converge to the optimum. To overcome this problem, Judice and Faustino [1992] reformulated this algorithm and solved it by first solving a sequence of linear complementary problems, which is formulated by treating the objective function as another constraint. Fortuny-Amat and McCarl [1981], again based on the KKT condition, formulated a much larger mixed-integer problem. The branch-and-bound algorithm such as that due to Bard and Falk [1982] is based on a series of transformations through the complimentary slackness conditions to transform the product terms into a series of equalities without altering the solution. The branch-and-bound technique is then applied to partition the feasible region and thus obtain a global optimal solution. Later, Bard and Moore [1990] modified the algorithm with 0-1 variables and extend it to solve quadratic BLPPs, which will be discussed in a later chapter for non-linear multiple-level problems. They claim that the performance and the robustness of the resulting algorithm are superior to all other contenders. However,

Ben-Ayed [1993] showed that, for large practical problems, the efficiency of the algorithm is still constrained by the exponential growth of the branch-and-bound tree. Hansen *et al.* [1992] formed a new branch-and-bound algorithm based on the tightness of the follower's constraint and used a mixed-integer approach.

To date, the most efficient traditional approaches appear to be algorithms based on branch-and-bound and the complementary-pivot concept such as these due to Bard and Moore [1990], Judice and Faustino [1992], and Hansen *et al.* [1992].

2.2 Extreme-point Search

As has already mentioned above, the extreme-point-search approach is not very efficient numerically. However, it does give a very good picture of the problems involved. Thus, we shall give a fairly detailed discussion with the help of an example and graphical illustrations.

Extreme-point search includes the kth-best algorithm and the grid-point-search method. Although some reviews [Kolstad 1985, Anandalingam and Friesz 1992] classify the latter as KKT methods, it seems more suitable to be classified as vertex enumeration [Wen and Hsu 1991].

2.2.1 k^{th}-best Algorithm

Assuming the feasible region is bounded and a unique solution exists for the lower-level problem for any feasible x_1, Bialas and Karwan [1982, 1984] proposed the kth-best algorithm with the following procedure:

Step 1. Start with $i = 1$ and solve Equation 2.1 with the given constraint set X by the simplex algorithm. Let this solution for the upper-level DM be represented by $x_{[1]}*$, and the possible solution set S = $\{x_{[i]}*\}$. Set the index set $I = \varnothing$. Go to Step 2.

Step 2. Solve Equation 2.3 with $x_1 = x_{[i]}*$. Let x^+ represents the optimal solution of this lower-level DM. If $x^+ = x_{[i]}*$, then stop and $x_{[i]}*$ is the global optimum with iteration index $k = i$. Otherwise, go to Step 3.

Step 3. Let $S_{[i]}$ denote the set of extreme points x which are adjacent to x^+ and such that $c_1^T x \leq c_1^T x_{[i]}*$. Let $I = I \cup (x_{[i]}*)$and $S = (S \cup S_{[i]}) \cap I^C$. Go to Step 4.

Step 4. Set $i = i+1$ and choose $x_{[i]}*$ so that $c_1 x_{[i]}* = \max_{x \in S_x}(c_1 x)$. Go to Step 2.

The search starts at the optimum of the upper-level problem. An overall optimal solution is reached if the upper-level optimum matches the lower-level one. Otherwise, search for the neighboring corners or extreme points of the previous point until the upper-level DM's proposed decision match the lower-level DM's optimum. From this iterative algorithm, we can see that the upper-level DM widens her tolerance or decision space continuously so that an overall optimum can be obtained. Thus, an implicitly compromise process is continuous carried out with each iteration.

To examine in more detail about this comprising process, let us consider the situation where, at the ith iteration, the solutions for the two problems are located at two neighboring vertices. The kth-best algorithm will force its solution to be either one of the two, depending on who moves first. The DM who moves first dominates the solution. Indeed this situation happens in all the extreme-point search procedures. The fundamental assumption of the kth-best algorithm is that the optimum should exist at the corner points. Since it is essentially impossible to define true optimality for multiple-objective problems, compromise or cooperation is usually needed in order to reach a solution. Instead of moving to one of the adjacent corner points, a more averaging compromise between these two extreme points would be more reasonable. In other words, instead of only one DM make the sacrifice, the sacrifice should be shared by both DMs. We shall see later that the proposed fuzzy approach comprises between both DMs and thus the results are more reasonable and, furthermore, the solution is not necessarily at the corner point.

The following example demonstrates the computational procedure of the kth-best algrithm

Example 2.1. A bi-level programming problem for balancing between trade surplus and profit [Zimmermann].

$$\max_{x1} f_1 = 2x_1 - x_2 \qquad \text{(trade surplus, upper level)}$$

where x_2 solves:

$$\max_{x2} f_2 = x_1 + 2x_2 \qquad \text{(profit, lower level)}$$

subject to:

$$\begin{array}{ll} 3x_1 - 5x_2 \le 15 & \text{(capacity)} \\ 3x_1 - x_2 \le 21 & \text{(management)} \\ 3x_1 + x_2 \le 27 & \text{(space)} \\ 3x_1 + 4x_2 \le 45 & \text{(material)} \\ x_1 + 3x_2 \le 30 & \text{(labor)} \\ x_1, x_2 \ge 0 & \end{array}$$

For simplicity, X will be used to represent the above constraint set in the following discussions. The actual solution procedure, based on the kth-best algorithm, can be summarized as:

Step 1. Solve the following upper-level problem:

$$\max f_1 = 2x_1 - x_2$$

subject to:

$$x \in X$$

The solution is $x_{[1]}$*= (7.5, 1.5) and is at vertex B (see Figure 2.1). Let the possible solution set S = {(7.5, 1.5)} and set the index set $I = \emptyset$. Go to Step 2.

Step 2. Solve the following lower-level problem with the results just obtained for the upper level:

$$\max f_2 = x_1 + 2x_2$$

subject to:

$$x \in X$$

with $x_1 = 7.5$. The solution for this problem is x^+= (7.5, 4.5). Since $x^+ \neq x_{[1]}*$, go to Step 3.

Step 3. Form the neighboring vertex set $S_{[1]}$= {(8, 3), (5, 0)}, which includes neighboring vertices A and C (see Figure 2.1). Let $I = I \cup \{x_{[1]}*\}$ = {(7.5, 1.5)}. The possible solution set is $S = (S \cup S_{[1]}) \cap I^C$ = {(8, 3), (5, 0)}. Go to Step 4.

Step 4. Set $i = i+1 = 2$, and choose $x_{[2]}* = \max_{x \in S}\{c_1\, x\}$= (8, 3). Thus the current solution is at vertex C. Go back to Step 2.

Step 2. Solve the lower-level problem again with $x_1 = 8$. The solution is x^+= (8, 3). Since $x^+ = x_{[2]}*$, the procedure terminates and $x_{[2]}*$ is the global optimum of the bi-level system with index $k = i = 2$.

Thus, the optimum obtained is x = (8, 3), f = (13, 14), which is at vertex C. The decision and objective spaces are shown in Figures 2.1 and 2.2, respectively. The values of the decision variables and the objective functions at the various vertexes are summarized in Table 2.1.

The example can also be treated as a Stackelberg game. The pay-off table or the bi-matrix form of the Stackelberg game is shown in Table 2.2. In addition to the vertexes and their values, the iteration sequences are also shown in parentheses. The moves in the game are searching sequences for the desired solution. The leader or upper-level DM as well as the follower or lower-level DM interacts with each other throughout the decision process as in Stackelberg game. This interaction terminates when the leader can no longer maximize his objective function. In such a situation, the Stackelberg equilibrium is reached. Since our problem has very few corner points, the matrix is very sparsed as is shown in Table 2.2

2.2.2 Grid-search Algorithm

Bard [1982, 1983b] proposed an algorithm based on sensitivity analysis to solve the linear two-stage optimization problem. The underlying theory rests on a set of first

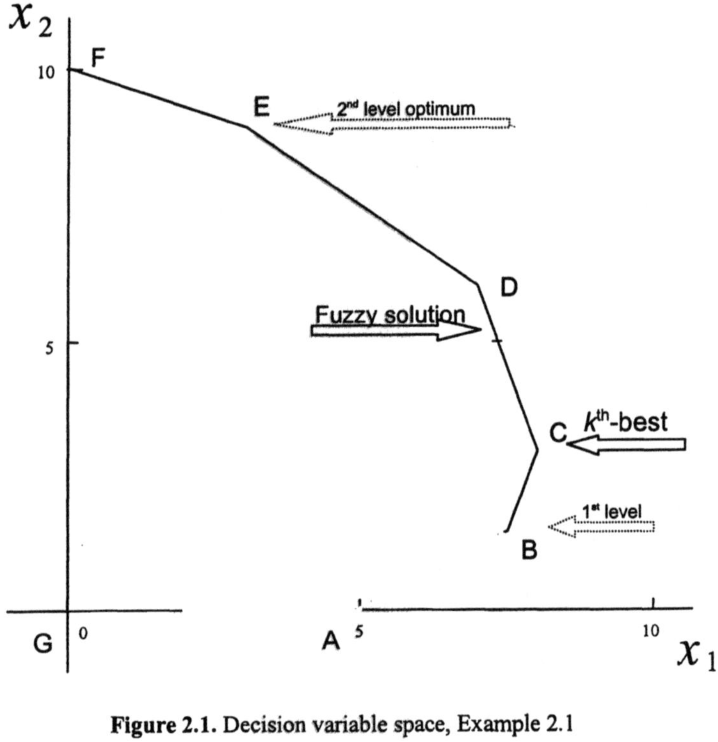

Figure 2.1. Decision variable space, Example 2.1

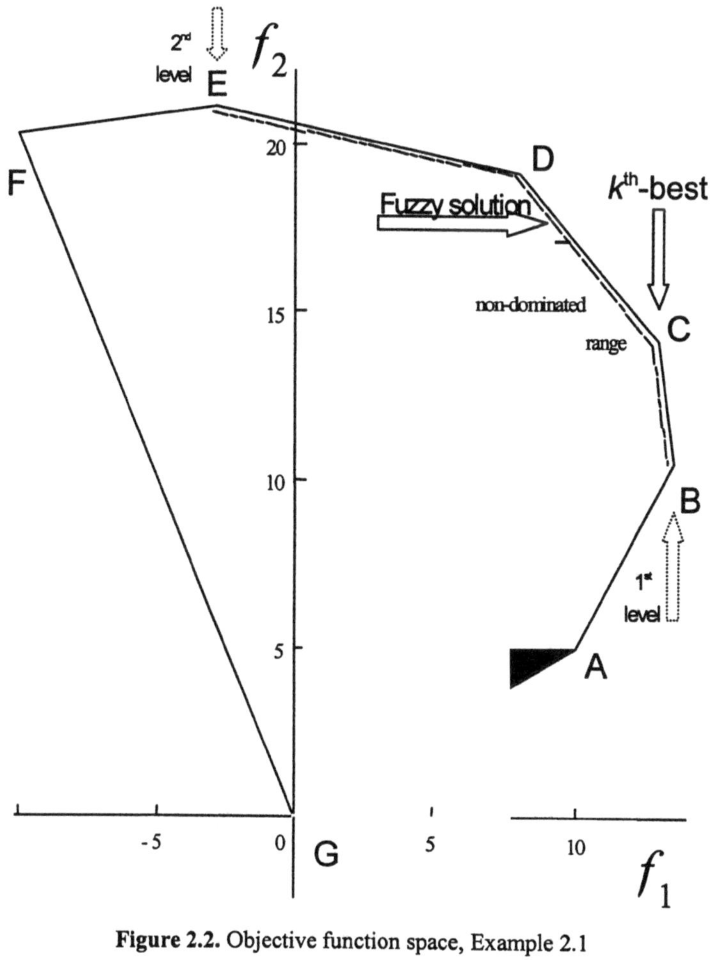

Figure 2.2. Objective function space, Example 2.1

Table 2.1. Numerical values at the various vertices, Example 2.1

Vertex	A	B	C	D	E	F	G
(x_1, x_2)	(5, 0)	(7.5, 1.5)	(8, 3)	(7, 6)	(3, 9)	(0, 10)	(0, 0)
(f_1, f_2)	(10, 5)	(13.5, 10.5)	(13, 14)	(8, 19)	(-3, 21)	(-10, 20)	(0, 0)

Table 2.2. Bi-matrix form, Example 2.1

X_1 \ X_2	0	1.5	3	6	9	10
0	G (0, 0)	-	-	-	-	F (-10, 20)
3		-	-	-	E (-3, 21)	-
5	A (10, 5)	-	-	-	-	-
7	-	-	-	D (8, 19)	-	-
7.5	-	B, (1) (13.5, 10.5)	-	-	-	-
8	-	-	C, (2) (13, 14)	-	-	-

order optimality conditions that parallel to the KKT conditions associated with a one-level parameterized linear program. By systematically varying the parameter over the unit interval and solving the corresponding linear program, the solution to the original problem is eventually obtained. The transformed parameterized expression can be expressed as:

$$\max P(\lambda', x_1, x_2) = \lambda'(c_{11}{}^{\mathrm{T}} x_1 + c_{12}{}^{\mathrm{T}} x_2) + (1-\lambda')\, c_{22}{}^{\mathrm{T}} x_2 \tag{2.6}$$

subject to:

$$A_1 x_1 + A_2 x_2 \le b$$
$$x_1, x_2 \ge 0$$
$$\lambda' \in [0, 1]$$

Again, in the following discussions, X represents the constraint set.

Once x_1 is obtained, the next step is to check the feasibility by solving the following inside problem:

$$\max f_2(x_2) = c_{22}{}^{\mathrm{T}} x_2 \tag{2.7}$$

subject to:

$$A_2 x_2 \leq b - A_1 \underline{x}_1$$
$$x_2 \geq 0$$

where $\underline{x}_1$ was obtained by solving Equation 2.6.

Bard [1982, 1983b] argued that there exists an $\lambda'^* \in [0, 1]$ such that the corresponding solution (x_1^*, x_2^*) of Equation 2.6 is both feasible and optimal to the original bi-level programming problem. Under certain non-degeneracy assumptions, the proposed algorithm can be described as:

Step 1. Let $\lambda' = 1$ and solve Equation 2.6 to obtain (x_1^1, x_2^1).

Step 2. Check (x_1^1, x_2^1) for feasibility by solving the inside problem, Equation 2.7, to obtain $\underline{x}_2^1$. If $\underline{x}_2^1 = x_2^1$, stop. Otherwise, go to Step 3.

Step 3. Let (x_1^k, x_2^k) be the current solution of Equation 2.6 with $\lambda' = \lambda'_k$. From sensitivity analysis, determine the minimum value of λ' such that (x_1^k, x_2^k) remains optimal. Call this value λ'_{min} and put $\lambda'_{k+1} = \lambda'_{min} - \delta$ where $\delta > 0$ is sufficiently small so that no vertices are missed.

Step 4. Solve Equation 2.6 for $P(\lambda'_{k+1}, x_1, x_2)$ to obtain an improved (x_1^{k+1}, x_2^{k+1}).

Step 5. Check (x_1^{k+1}, x_2^{k+1}) for feasibility by solving the inside problem, Equation 2.7, to obtain $\underline{x}_2^{k+1}$. If $\underline{x}_2^{k+1} = x_2^{k+1}$, stop. Otherwise, let $k = k + 1$ and go to Step 3.

Notice that there may have multiple optimal solutions in Step 4 if x_2 with a given x_1 is not a singleton. In that case, the algorithm must iterate for each of these vertices. Let us illustrate the solution procedure with the following example:

Example 2.2. Consider the linear problem [Bard and Falk 1982]:

$$\max_x f_1 = 2x_1 - x_2 - 0.5y_1$$

where x solves:

$$\max_y f_2 = -x_1 - x_2 + 4y_1 - y_2$$

subject to:

$$-2x_1 + y_1 - y_2 \leq -2.5$$
$$x_1 - 3x_2 + y_2 \leq 2$$
$$2x_1 + x_2 \leq 2$$
$$x, y \geq 0$$

The parameterized linear program for the above problem is:

$$\max P(\lambda', x, y) = 2\lambda' x_1 - \lambda' x_2 + (4 - 4.5\lambda') y_1 + (1 - \lambda') y_2$$

subject to:

$$(x, y) \in X$$

where X represents the set of the original constraints.

The above problem was solved. The iteration results for the grid search are summarized in Table 2.3. Feasibility is determined by solving the following inside problem:

$$\max f_2 = 4y_1 - y_2$$

subject to:

$$y_1 - y_2 \le -2.5 + 2x_1^k$$
$$y_2 \le 2 - x_1^k + 3x_2^k$$
$$y \ge 0$$

with the values for (x_1^k, x_2^k) obtained from the above parameterized linear program. As shown in Table 2.3, the objective of the upper level decreases as λ' decreases. The global optimum is obtained at the first feasible point. The optimum is $(x_1, x_2, y_1, y_2) = (1, 0, 0.5, 1)$ and the objectives are $f = (1.75, -1)$ for the upper and the lower levels, respectively. For purpose of illustration, the computation is continued until λ' is reduced to zero.

Table 2.3. Iteration results for Example 2.2.

λ'	(x_1, x_2, y_1, y_2)	f_1	Feasibility
1.0-0.98	(1, 0, 0, 1)	2.0	No
0.97-0.84	(1, 0, 0, 0.5)	2.0	No
0.83-0.67	(1, 0, 0.5, 1)	1.75	Yes
0.66-0	(0, 2, 5.5, 8)	-4.75	Yes

Table 2.4. Comparison of algorithms (from Bard [1983b])

Algorithm	Problem size (n_1, n_2, m)	Average CPU time	Average number of iterations	Range of iterations
Branch-and-bound (Bard and Falk)	(6,3,7) (10,10,6) (30,20,25)	77 136 215	321 786 1142	106-630 340-1010 510-1570
Implicit search (Candler and Townsley)	(6,3,7) (10,10,16) (30,20,25)	34 80 194	63 107 232	53-88 47-191 126-430
PCP algorithm (Bialas *et al.*)	(6,3,7) (10,10,16) (30,20,25)	2 5 9	12 21 47	7-18 14-26 13-123
k^{th}-best (Bialas and Karwan)	(6,3,7) (10,10,16) (30,20,25)	4 18 66	77 229 966	61-109 153-576 423-2011
Grid search	(6,3,7) (10,10,16) (30,20,25)	1 1 1	5 11 23	3-8 4-16 11-38

The grid-search algorithm is designed to solve Equation 2.6 parametrically and to terminate when the first feasible point is reached. Bard's [1983b] computational results are reproduced in Table 2.4. The symbols n_1 and n_2 are the number of decision variables for the upper and lower levels, respectively, and m represents the number of constraints. The CPU time listed for comparison purposes only. As can be seen from this table, grid-search algorithm has an overall advantage over the algorithms of the kth-best approach [Bialas and Karwan 1982, 1984], parametric complementary-pivot (PCP) algorithm [Bialas and Karwan 1982], implicit search [Candler and Townsley 1982], and the branch-and-bound algorithm [Bard and Falk 1982]. However, Haurie *et al.* [1990] and Ben-Ayed and Blair [1990] have shown that this algorithm may fail to obtain the optimum.

2.3 Transformation Approach

As discussed before, the transformation approach uses the KKT optimality condition or other similar functions to convert the lower-level problem into the constraint of the upper level. The transformed problem such as Equation 2.5 is one less level than the original two-level one. Thus, this nested optimization problem is reduced to the traditional one-level non-linear programming problem, which is non-convex. By using the traditional optimization approaches such as branch-and-bound and penalty-function, various algorithms have been developed. Some of them appear to be the most efficient approaches among the traditional optimization techniques for this duo-ploy type problem.

2.3.1 Mixed-integer Approach

Fortuny-Amat and McCarl [1981] developed a mixed-integer approach by considering the disjunctive nature of the complementary slackness of the non-linear term in the transformed equation. Because of the introduction of the zero-one variables, the resulting formulation forms a much larger programming problem. However, the basic idea is a very useful one and has been incorporated into several algorithms. Notice that in the transformed problem, Equation 2.5, the only non-linear equation is the complimentary slackness equation:

$$w^T (A_1 x_1 + A_2 x_2 - b) = 0$$

If we let:

$$y = A_1 x_1 + A_2 x_2 - b$$

then, we obtain $w^T y = 0$, which is consisted of the sums of the products of two variables. In order to satisfy this equation, one of the two variables, either w_i or y_i, must be equal to zero. This condition can be achieved by introducing the zero-one variable η for each product. Thus, we have:

$$w \leq (1-\eta) M$$

$$y = A_1 x_1 + A_2 x_2 - b \leq M\eta$$

where M is a very large positive constant. Using the above two equations, Equation 2.5 can be restated as:

$$\max f_1 (x_1, x_2) = c_{11}^{\mathrm{T}} x_1 + c_{12}^{\mathrm{T}} x_2 \tag{2.8}$$

subject to:

$$\begin{aligned}
&A_1 x_1 + A_2 x_2 \leq b \\
&w \leq (1\text{-}\eta) M \\
&A_1 x_1 + A_2 \mathrm{x}_2 - b \leq M\eta \\
&\mathrm{w}^{\mathrm{T}} \mathrm{A}_2 = c_{22} \\
&\eta \in \{0, 1\} \\
&x_1, \mathrm{x}_2, w \geq 0
\end{aligned}$$

Compared to Equation 2.5, the non-linear term disappeared and is replaced by two linear inequalities. Due to the presence of the zero-one integer, the problem represented by Equation 2.8 is a mixed-integer problem.

To illustrate the algorithm, the numerical example solved in Example 2.1 is again solved by this mixed-integer approach in the following.

Example 2.3. Using Equation 2.8, the numerical example used in Example 2.1 becomes:

$$\max f_1 = 2x_1 - x_2$$

subject to:

$$\begin{aligned}
&3x_1 - 5x_2 \leq 15 \\
&3x_1 - x_2 \leq 21 \\
&3x_1 + x_2 \leq 27 \\
&3x_1 + 4x_2 \leq 45 \\
&\mathrm{x}_1 + 3x_2 \leq 30 \\
&-3\mathrm{x}_1 + 5x_2 + 15 \leq M\eta_1, \quad \mathrm{w}_1 \leq (1-\eta_1)M \\
&-3x_1 + x_2 + 21 \leq M\eta_2, \quad w_2 \leq (1-\eta_2)M \\
&-3x_1 - x_2 + 27 \leq M\eta_3, \quad w_3 \leq (1-\eta_3)M \\
&-3x_1 - 4x_2 + 45 \leq M\eta_4, \quad w_4 \leq (1-\eta_4)M \\
&-x_1 - 3x_2 + 30 \leq M\eta_5, \quad w_5 \leq (1-\eta_5)M \\
&-5w_1 - w_2 + w_3 + 4w_4 + 3w_5 = 2 \\
&\eta_1, \eta_2, \eta_3, \eta_4, \eta_5 \in \{0, 1\} \\
&\mathrm{w}_1, w_2, w_3, w_4, w_5, \mathrm{x}_1, x_2 \geq 0
\end{aligned}$$

Again, M is a very large positive constant.

The LINGO programming code [1992] was used to solve this problem. With the large constant $M = 10{,}000$, the solution obtained is $x = (x_1, x_2) = (7.5, 1.5)$, $f_1 = 13.5$, and $f_2 = 10.5$. The optimum is at vertex B (see Figures 2.1 and 2.2). We also solved the problem with M = 100,000, 1,000,000, 10,000,000, and 100,000,000, the solution remains the same. The solution is quite stable and is different from that obtained by the kth-best approach, whose solution optimum was at point C. It appears that the upper-level DM dominates the solution for this particular case.

Because of the introduction of the zero-one integer variables, η, this mixed-integer programming problem is not easy to solve. This is especially true for practical problems, which are usually large and complex. However, the concept has been used to form a branch-and-bound tree, which forms the basis for fairly effective branch-and-bound algorithms.

Fortuny-Amat and McCarl [1981] formulated the problem as a two-level quadratic programming problem and used the distributor and customer as a two-level example. In particular, these investigators considered the problem of purchasing fertilizer from two different sources, the local neighborhood versus competitors from other places.

2.3.2 Complementary-pivot Algorithm

As has been pointed out previously, the complementary-pivot algorithm, introduced by Bialas and Karwan [1984] may not converge to the optimum. However, because the approach forms the basis of later developments, the original algorithm is discussed in the following.

2.3.2.1 Parametric Complementary-pivot Algorithm

Bialas and Karwan [1984] proposed the parametric complementary-pivot (PCP) algorithm to solve the BLPP problem. The algorithm uses Wolfe's approach for quadratic programming [1959] and transforms the objective function into the constraint, $c_{11}^T x_1 + c_{22}^T x_2 \geq \alpha'$, where α' is a constant. At the beginning of the iteration, a small value for α' should be used. As the computation progresses, the value of α' should be increased gradually and parametrically to improve the upper-level's objective. At the same time, the non-linear complementary slackness terms are taken care of by the use of restricted entry into the basis. Our discussion will follow the work of Bialas and Karwan [1984].

The transformed Equation 2.5 can be rewritten in the following complementary slackness form:

$$\max_{x1,x2} f_1 (x_1, x_2) = c_{11}{}^T x_1 + c_{12}{}^T x_2$$

subject to:

$$\begin{aligned}
&A_1 x_1 + A_2 x_2 + u = b \\
&w^T A_2 - v = c_{22} \\
&w^T u = 0 \\
&x_2 v = 0
\end{aligned}$$

$$x_1, x_2, v, u, w \geq 0$$

We have introduced the complimentary variables u and v. Instead of solving the above problem directly, the PCP algorithm considers the following system of equations [Bialas and Karwan 1984]:

$$\begin{aligned}
&A_1 x_1 + A_2 x_2 + u = b \\
&-\varepsilon H x_1 + w^T A_2 - v = c_{22} \\
&c_{11}{}^T x_1 + c_{12}{}^T x_2 \geq \alpha' \\
&w\,u = x_2\,v = 0 \\
&x_1, x_2, u, v, w \geq 0
\end{aligned}$$

The maximization expression has been replaced by the third inequality in the above equations. The matrix H may be any negative definite matrix, and ε is a suitably small positive scalar so that the term εH leads only to a small perturbation of Equation 2.5. Bialas and Karwan suggested a value of $H = -\mathrm{I}$ with ε ranging from 10^{-2} to 10^{-8}.

The PCP algorithm can be viewed as an implicit enumeration of the lower-level optimal bases. All of the computations for the PCP algorithm are performed within the framework of the simplex-like tableau whose size is roughly that of the original system. In the absence of degeneracy and multiple level-two optima, a simplified version of the PCP algorithm can be described as [Bialas and Karwan 1984]:

Step 0. Let $\alpha = \alpha_{min}$, where α_{min} is a lower bound for $c_{11}{}^T x_1 + c_{12}{}^T x_2$.

Step 1. Obtain a feasible solution for:

$$\begin{aligned}
&A_1 x_1 + A_2 x_2 + u = b \\
&-c_{21}{}^T x_1 - c_{22}{}^T x_2 + s = -\alpha' \\
&-\varepsilon H x_2 + w^T A_2 - v = c_{22}
\end{aligned} \tag{2.9}$$

by using the Wolfe's two-phase quadratic programming algorithm. The non-linear expressions, wu and x_2v, can be satisfied by controlling the variables entering the basis. In other words, use the following conditions for variables entering the basis:

1.1. if x_{2k} is in the basis and is not the leaving variable for the current pivot, do not admit v_k into the basis, and *vice versa*, and

1.2. if u_k is in the basis and is not the leaving variable for the current pivot, do not admit w_k into the basis, and *vice versa*.

If a feasible solution exists, go to Step 2. Otherwise, go to Step 3.

Step 2. Let $x^* = (x_1{}^*, x_2{}^*)$ represents the incumbent solution and B represents the current basis. Let $\beta = (b, -\alpha', c_{22})$, represents the right-hand side vector in Equation 2.9. Define $\underline{\beta} = B^{-1}\beta$ and let the column corresponding to s in the simplex tableau be the kth column denoted by y_k. Choose $\Delta\alpha'$ = min

$\{\beta_i / y_{ki} \mid y_{ki} > 0\}$. Let $\alpha' = (\alpha' + \Delta\alpha')(1 + \gamma)$ where γ is a suitably small positive scalar. Go to Step 1.

Step 3. Stop, if x^* is the solution which is feasible within the constrained space of f_2 and $c_{21}^T x_1^* + c_{22}^T x_2^*$ is within $\gamma \times 100\%$ of the optimal objective function value.

To illustrate the procedure, the example solved by Bialas and Karwan [1984] is used in the following.

Example 2.4. Bialas and Karwan [1984] solved the following two-level linear resource control problem. We follow their numerical discussions.

$$\max_{x1} f_1 = x_2$$

where x_2 solves:

$$\max_{x2} f_2 = -x_2$$

subject to:

$$\begin{aligned}
&-x_1 - 2x_2 \le -10 \\
&x_1 - 2x_2 \le 6 \\
&2x_1 - x_2 \le 21 \\
&x_1 + 2x_2 \le 38 \\
&-x_1 + 2x_2 \le 18 \\
&x_1, x_2 \ge 0
\end{aligned}$$

Following the above outlined procedure, the PCP algorithm deals with the following transformed expression:

$$\max f_1 = x_2$$

subject to:

$$\begin{aligned}
&-x_1 - 2x_2 + u_1 = -10 \\
&x_1 - 2x_2 + u_2 = 6 \\
&2x_1 - x_2 + u_3 = 21 \\
&x_1 + 2x_2 + u_4 = 38 \\
&-x_1 + 2x_2 + u_5 = 18 \\
&-x_2 + s = -\alpha' \\
&-\varepsilon x_2 - 2w_1 - 2w_2 - w_3 + 2w_4 + 2w_5 - v_1 = -1 \\
&u_1 w_1 = u_2 w_2 = u_3 w_3 = u_4 w_4 = u_5 w_5 = 0 \\
&x_1, x_2, u_1, u_2, u_3, u_4, u_5, s, w_1, w_2, w_3, w_4, w_5, v_1 \ge 0
\end{aligned}$$

Let $\varepsilon = 10^{-8}$. Since $f_1 = x_2 \ge 0$, choose $\alpha'_{\min} = 0$ to start the iteration. In Step 1, consider the first phase of the two-phase simplex. In other words, with $\alpha' = 0$, minimize the sum of the artificial variables:

$$\min g_1 = u_1 + s + v_1$$

subject to:

$$\begin{aligned}
&x_1 + 2x_2 - u_1 = 10 \\
&x_1 - 2x_2 + u_2 = 6 \\
&2x_1 - x_2 + u_3 = 21 \\
&x_1 + 2x_2 + u_4 = 38 \\
&-x_1 + 2x_2 + u_5 = 18 \\
&-x_2 + s = 0 \\
&-10^{-8} x_2 - 2w_1 - 2w_2 - w_3 + 2w_4 + 2w_5 - v_1 = -1 \\
&x_1, x_2, u_1, u_2, u_3, u_4, u_5, s, w_1, w_2, w_3, w_4, w_5, v_1 \geq 0
\end{aligned}$$

The solution for this problem is:

$$\begin{aligned}
&x = (x_1, x_2) = (8, 1) \\
&u = (u_1, u_2, u_3, u_4, u_5) = (0, 0, 6, 28, 24) \\
&w = (w_1, w_2, w_3, w_4, w_5) = (0.5, 0, 0, 0, 0) \\
&s = 1 \\
&v_1 = 0
\end{aligned}$$

In Step 2, let $\gamma = 0.01$ and $\Delta\alpha = 1$, the value of α is set $(\alpha' + \Delta\alpha')(1 + 0.01) = 1.01$. Re-solving the above expression with the new values, obtain the solution as:

$$\begin{aligned}
&x = (12, 3) \\
&u = (8, 0, 0, 20, 24) \\
&w = (0, 0.5, 0, 0, 0) \\
&s = 1.99 \\
&v_1 = 0
\end{aligned}$$

Computing $\Delta\alpha' = 1.99$ and resetting α' equal to $(1.01 + 1.99)(1.01) = 3.03$, solve the above expression again and the solution obtained is:

$$\begin{aligned}
&x = (16, 11) \\
&u = (28, 12, 0, 0, 12) \\
&w = (0, 0, 1, 0, 0) \\
&s = 7.97 \\
&v_1 = 0
\end{aligned}$$

Compute $\Delta\alpha' = 7.97$ and set $\alpha' = 11.11$. Return to Step 1, there is no solution existed with $\alpha' = 11.11$. Hence, the solution $x^* = (16, 11)$ is the solution of the two-level system.

Bialas and Karwan [1984] carried out some computational experiments to test the effectiveness of both the kth-best and the PCP algorithms. The number of variables controlled by the second level appears to have more influence on the performances and the problem is easier to solve with fewer level-two variables. Furthermore, it appears that the kth-best algorithm can be erratic under certain conditions and fails to obtain a solution within a reasonable computational time. Thus, they concluded that the PCP algorithm is a more reliable approach.

2.3.2.2 Sequential Linear Complementary Algorithm with Branch-and-bound

The PCP algorithm due to Bialas and Karwan [1984] sometimes does not converge and is unable to solve the BLPP in all cases. To overcome this problem, Judice and Fausstino [1992] modified and reformulated the problem and this modified problem is known as the sequential linear complementary problem (SLCP). Instead of using the original algorithm, these investigators, with the help of a branch-and-bound search, solved a sequence of linear complementary problems, which converge to the desired optimum of the original problem. These authors also carried out a fairly extensive numerical study, and found out that, compared to the branch-and-bound algorithm due to Bard and Moore [1990], the proposed approach was consistently more efficient. Furthermore, the current approach appears to be more advantageous as the dimension of the problem increases. Another advantage is that the algorithm is robust and is fairly efficient in finding a global optimum.

By adding complementary variables, Equation 2.5 can be rewritten as:

$$\max_{x1,x2} f_1(x_1, x_2) = c_{11}^{\mathrm{T}} x_1 + c_{12}^{\mathrm{T}} x_2 \tag{2.10}$$

subject to:

$$\begin{aligned}
&A_1 x_1 + A_2 x_2 + u = b \\
&w^{\mathrm{T}} A_2 - v = c_{22} \\
&w^{\mathrm{T}} u = 0 \\
&x_2 v = 0 \\
&x_1, x_2, v, u, w \geq 0
\end{aligned}$$

The last two equations before the non-negativity conditions are complementary slackness conditions. Judice and Faustino introduced a parameter λ and replaced the objective function by the constraint:

$$c_{11}^{\mathrm{T}} x_1 + c_{12}^{\mathrm{T}} x_2 \geq \lambda$$

After introducing the slack variable, β, the linear complementary problem becomes:

$$\begin{aligned}
&A_1 x_1 + A_2 x_2 + u = b \\
&A_2^{\mathrm{T}} w - v = c_{22} \\
&c_{11}^{\mathrm{T}} x_1 + c_{12}^{\mathrm{T}} x_2 + \beta = \lambda \\
&w^{\mathrm{T}} u = x_2 v = 0 \\
&x_1, x_2, v, u, w \geq 0
\end{aligned}$$

which can be put in matrix form:

$$\begin{bmatrix} -u \\ -v \\ \beta \end{bmatrix} = \begin{bmatrix} -b \\ c_{22} \\ 0 \end{bmatrix} + \begin{bmatrix} 0 \\ 0 \\ 1 \end{bmatrix} \lambda + \begin{bmatrix} 0 & A_1 & A_2 \\ -A_2^T & 0 & 0 \\ 0 & -c_{11}^T & -c_{12}^T \end{bmatrix} \cdot \begin{bmatrix} w \\ x_1 \\ x_2 \end{bmatrix} \tag{2.11}$$

$$w^T u = x_2 v = 0$$
$$x_1, x_2, v, u, w \geq 0$$

The efficiency of the proposed algorithm depends essentially on the solution of the linear complementary problem, Equation 2.11. Judice and Faustino [1992] developed a hybrid enumerative tree-search technique to solve this linear complementary problem (LCP). These investigators further improved the efficiency by exploring various strategies to reduce the search. The hybrid enumerative approach is essentially a branch-and bound enumerative technique, whose efficiency is improved by the incorporation of a modified reduced gradient method due to Al-Khayyal [1987]. Equation 2.11 can be represented by the following vector equation:

$$x = q + My + Nz \tag{2.12a}$$
$$x \geq 0, y \geq 0, z \geq 0 \tag{2.12b}$$
$$x_I y_I = 0, I = 1, 2, 3, \ldots, n_2 + m \tag{2.12c}$$

where x and y are $(n_1 + m + 1)$- and $(n_1 + m)$-dimensional vectors, respectively; z is an m-dimensional vector; and q is an $(n_1 + m + 1)$-dimensional vector. Following the tradition of linear programming, a solution of Equation 2.12a satisfying the constraints of Equation 2.12b is a feasible solution. A complementary solution (x, y, z) must satisfy Equations 2.12a-2.12c. The enumerative branch-and-bound method attempts to find a complementary solution by using only basic feasible solutions of Equation 2.12a. This is accomplished by exploring the branch-and–bound tree shown in Figure 2.3 [Judice and Faustino 1992], where $i_1, i_2, \ldots$ are integer numbers.

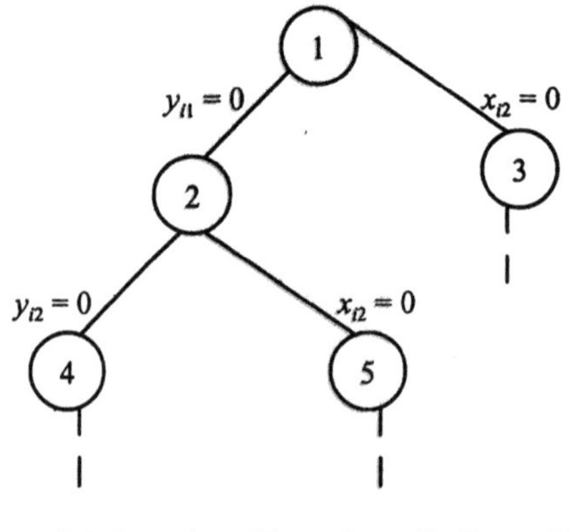

Figure 2.3. Branch-and-bound tree [Judice and Faustino 1992]

To begin the process, an initial feasible solution for node 1 is obtained by solving the problem:

$$\min y_0 \tag{2.13}$$

subject to:

$$x = q + p\,y_0 + M\,y + N\,z$$
$$x \geq 0,\ y \geq 0,\ z \geq 0,\ y_0 \geq 0$$

where y_0 is an artificial variable and p is a non-negative vector satisfying $p_i > 0$ for all i such that $q_i < 0$, which essentially follows the rule used by Wolfe [1959]. This problem is solved by using the simplex algorithm with a modification that whenever possible Equation 2.12c is also satisfied. The remaining notes are generated in essentially the same manner, namely, solving a problem similar to Equation 2.13 except minimizing either y_{ik} or x_{ik} with k = 2, 3,..., and with the variables in earlier notes kept to zero. Two situations may occur during the generation of any one of the notes:

1. if the minimum of zero is obtained, then it is fixed at zero and continues to generate the next generation of notes at the descending path of the tree;
2. if the minimum value is positive, then this branch is pruned and the node is fathomed.

This enumerative branching is continued until either a complementary solution is obtained or the problem has no solution. The algorithm can be summarized as [Judice and Faustino 1992]:

Step 1. Obtain an initial feasible solution. If no feasible solution can be obtained, the problem has no solution. Otherwise, go to Step 2.

Step 2. If the solution obtained satisfies Equation 2.12c, or complementary; then a solution is found and stop. Otherwise, go to Step 3.

Step 3. Choose two sets of complementary positive basic variables x_i and y_i and generate two nodes by solving two linear programming problems, which minimize x_i or y_i subject to the linear constraints with x_i and y_i at earlier nodes set to zero. If the minimum of the variable is positive, then the corresponding node is fathomed.

Step 4. If all the nodes are fathomed, then the problem has no solution and stop. Otherwise, choose an unfathomed node and got to 2.

In order to carry out the above procedure, the following details are needed: the choice of the initial feasible solution in Step 1, the choice of the two sets of the complementary variables x_i and y_i in Step 3, and the choice of the unfathomed node in Step 4. Obviously, these choices influence tremendously the efficiency of the algorithm. This is especially true if there are many nodes to be obtained along the branch-and-bound tree. Judice and Faustino [1992] incorporated a number of procedures to help the above choices and thus increase the search efficiency. One of the heuristics used is the reduced gradient search method due to Al-Khayyal [1987]. This gradient method can find a local optimum of the function:

$$f(x, y, z) = \sum_{i=1}^{m+n_2} x_i y_i$$

Fairly large computational experiments were carried out by Judice and Faustino [1992] and compared with other approaches. The maximum number of constraints, the maximum number of variables controlled by the leader, and the maximum number of variables controlled by the follower in these experiments are 150, 300, and 150, respectively. For small-dimensional problems, this improved hybrid enumerative SLCP approach is about par with the branch-and–bound approach due to Bard and Moore [1992]. However, for large-dimensional problems, the present approach is significantly more efficient then the branch-and-bound algorithm.

2.3.3 Branch-and-bound Algorithm

Among the many approaches for solving the BLPP problems, the branch-and-bound algorithm appears to be one of the more effective approaches among the traditional optimization techniques. Fortuny-Amat and McCarl [1981] suggested the use of two equality constraints with zero-one variables to replace the non-linear complementary term. Bard and Moore [1990] used this concept to improve their earlier branch-and-bound algorithm [Bard and Falk 1982] and the resulting formulation can be used to solve linear and quadratic BLPP. In the following discussion, we have restricted our discussions to the linear problem. Bard and Moore formulated their problem in the quadratic form, which will be discussed in the next chapter. Hansen *et al.* [1992] follows the concept used in global optimization, proposed a new algorithm by exploiting the follower's tight constraint. They did fairly extensive numerical studies and their algorithm appears to be the most efficient one among the proposed branch-and bound approaches.

2.3.3.1 Algorithm of Bard and Moore

To formulate the algorithm, let us rewrite Equation 2.5 to include the non-negativity constraints of the decision variables for the lower-level decision variable, x_2, explicitly:

$$\max_{x1,x2} f_1 (x_1, x_2) = c_{11}{}^{\mathrm{T}} x_1 + c_{12}{}^{\mathrm{T}} x_2 \qquad (2.14)$$

subject to:

$$A_1 x_1 + A_2 x_2 \leq b$$
$$w_1^{\mathrm{T}} (A_1 x_1 + A_2 x_2 - b) + w_2^{\mathrm{T}} x_2 = 0$$
$$w_1^{\mathrm{T}} A_2 + w_2^{\mathrm{T}} = c_{22}$$
$$x_1, x_2, w, u \geq 0$$

where $w_1 = 1,2,\ldots, m$, and $w_2 = 1,2,\ldots, n_2$, are dual vectors or Lagrange multipliers for the constraints of the original problem and the non-negativity constraints for the lower-level decision variables, respectively. Rewrite the above equations without the complementary non-linear equation, we have:

$$\max_{x1,x2} f_1 (x_1, x_2) = c_{11}^{\mathrm{T}} x_1 + c_{12}^{\mathrm{T}} x_2 \tag{2.15}$$

subject to:

$$A_1 x_1 + A_2 x_2 \leq b$$
$$\mathrm{w}_1^{\mathrm{T}} A_2 + w_2^{\mathrm{T}} = c_{22}$$
$$x_1, x_2, w \geq 0$$

and write the complementary equation separately:

$$w_1^{\mathrm{T}} (A_1 x_1 + A_2 x_2 - b) + w_2^{\mathrm{T}} x_2 = 0 \tag{2.16}$$

which can also be represented as:

$$w_i\, q_i = 0, \;\; \mathrm{i} = 1,2,\ldots, m{+}n_2 \tag{2.17}$$

where $w^{\mathrm{T}} = [w_1, w_2]$, the first m q's equal to the correspondence elements of $(A_1 x_1 + A_2 x_2 - b)$, and the last n_2 q's equal to the correspondence elements of the vector x_2. Notice that Equation 2.15 is linear and thus can be solved easily. The procedure is to solve Equation 2.15 iteratively. If the results for any iteration satisfy Equation 2.16 or 2.17 for all i, then the corresponding point is in the inducible region and thus is a potential solution for the BLPP. If not, branch-and-bound algorithm is used to examine implicitly all combinations of the complementary slackness, Equation 2.17.

To formulate the algorithm, Bard and Moore [1990] introduced some additional notation. Let $v = \{1,2,\ldots, m + n_2\}$ be the index set for the terms in Equation 2.17, and let $\underline{f}_1$ be the incumbent lower bound on the upper-level's objective. At the kth level of the branch-and-bound tree we define a subset of indices $v_k \subseteq v$, and a path vector P_k corresponding to an assignment of $w_i = 0$ or $q_i = 0$ for $i \in v_k$. Then, let:

$$S_k^+ = \{i \mid i \in v_k \text{ and } w_i = 0\}$$
$$S_k^- = \{i \mid i \in v_k \text{ and } q_i = 0\}$$
$$S_k^0 = \{i \mid \mathrm{i} \notin v_k\}$$

For $i \in S_k^0$, the variables w_i and q_i are free to assume any non-negative values in the solution of Equation 2.15 and thus Equation 2.17 will not necessarily be satisfied. The proposed procedure is summarized in the following [Bard and Moore 1990]:

Step 0. (Initialization) Let $k = 0$, $S_k^+ = \varnothing$, $S_k^- = \varnothing$, $S_k^0 = \{1, 2, \ldots, m + n_2\}$, and $\underline{f}_1 = -\infty$.

Step 1. (Iteration k) Set $w_i = 0$ for $i \in S_k^+$ and $q_i = 0$ for $i \in S_k^-$. Solve Equation 2.15. If the solution is infeasible, go to Step 5. Otherwise, let $k = k + 1$ and label the solution as (x_1^k, x_2^k, w^k). Go to Step 2.

Step 2. (Fathoming) If $f_1(x_1, x_2) \le \underline{f}_1$, go to Step 5. Otherwise, go to the next step.

Step 3. (Branching) If $w_i^k\ q_i\ (x_1^k, x_2^k) = 0$, $i = 1, 2, \ldots, m+n_2$, go to Step 4. Otherwise, select i for which $w_i^k\ q_i\ (x_1^k, x_2^k)$ is the largest and label it i_1. Let $S_k^+ \leftarrow S_k^+ \cup \{i_1\}$, $S_k^0 \leftarrow S_k^0 \setminus \{i_1\}$, $S_k^- \leftarrow S_k^-$, append i_1 to P_k, and go to Step 1.

Step 4. (Updating) $\underline{f}_1 = f_1(x_1^k, x_2^k)$. Go to the next step.

Step 5. (Backtracking) If no live node exists, go to Step 6. Otherwise, branch to the newest live vertex and update S_k^+, S_k^-, S_k^0 and P_k. Go to Step 1.

Step 6. (Termination) If $\underline{f}_1 = -\infty$, there is no feasible solution to the BLPP. Otherwise, declare the feasible point associated with $\underline{f}_1$ the optimal solution.

A live node is one associated with a sub-problem that has not yet been fathomed at either Step 1 due to unfeasibility or at Step 2 due to bounding.

To illustrate the algorithm, we follow the example used by Bard and Moore [1990] in the following.

Example 2.5. Consider the linear two-level problem [Candler and Townsley 1982, Bard and Moore 1990]:

$$\max_x f_1 = 8x_1 + 4x_2 - 4y_1 + 40y_2 + 4y_3$$

where x solves:

$$\max_y f_2 = -x_1 - 2x_2 - y_1 - y_2 - 2y_3$$

subject to:

$$\begin{aligned}
&y_1 - y_2 - y_3 \ge -1 \\
&-2x_1 + y_1 - 2y_2 + 0.5y_3 \ge -1 \\
&-2x_2 - 2y_1 + y_2 + 0.5y_3 \ge -1 \\
&x_1, x_2, y_1, y_2, y_3 \ge 0
\end{aligned}$$

The transformed problem without the complementary slackness product terms is:

$$\max f_1 = 8x_1 + 4x_2 - 4y_1 + 40y_2 + 4y_3$$

subject to:

$$
\begin{aligned}
&y_1 - y_2 - y_3 \geq -1\\
&-2x_1 + y_1 - 2y_2 + 0.5y_3 \geq -1\\
&-2x_2 - 2y_1 + y_2 + 0.5y_3 \geq -1\\
&-w_1 - w_2 + w_3 - w_4 = -1\\
&w_1 + 2w_2 - w_3 - w_5 = -1\\
&w_1 - 0.5w_2 - 0.5w_3 - w_6 = -2\\
&x_1, x_2, y_1, y_2, y_3, w_1, w_2, w_3, w_4, w_5, w_6 \geq 0
\end{aligned}
$$

and the complementary slackness product terms are:

$$
\begin{aligned}
&w_1\,(y_1 - y_2 - y_3 + 1) = 0\\
&w_2\,(-2x_1 + y_1 - 2y_2 + 0.5y_3 + 1) = 0\\
&w_3\,(-2x_2 - 2y_1 + y_2 + 0.5y_3 + 1) = 0\\
&w_4\,y_1 = 0\\
&w_5\,y_2 = 0\\
&w_6\,y_3 = 0
\end{aligned}
$$

As pointed out by Bard and Moore [1990], there are six of these product terms and thus, direct search approach needs to solve $2^7 - 1 = 127$ sub-problems. By using the proposed branch-and-bound approach, the optimal solution was reached by the fourth iteration and confirmed after ten sub-problems were examined.

The actual branch-and-bound tree for solving this problem is reproduced in Figure 2.4 from Bard and Moore [1990]. The initial feasible solution for the transformed problem without the complementary slackness product terms is:

$$
\begin{aligned}
&\boldsymbol{X} = (x_1, x_2) = (0, 0)\\
&\boldsymbol{Y} = (y_1, y_2, y_3) = (1.5, 1.5, 1)\\
&\boldsymbol{W} = (w_1, w_2, w_3, w_4, w_5, w_6) = (0, 0, 0, 1, 1, 2)\\
&F_1\,(\boldsymbol{x}, \boldsymbol{y}) = 58
\end{aligned}
$$

The above results should satisfy Step 2 since $\underline{f}_1$ is set at $-\infty$ at the initiation step. However, it does not satisfy Equation 2.17. Thus, at Step 3, select the highest value of the complementary slackness product term, which turn out to be $w_6 q_6 = 2$ with $i_1 = 6$. We obtain, at the first level, $S_1^+ = \{6\}$, $S_1^- = 0$, $S_1^0 = (1, 2, 3, 4, 5)$, $P_1 = 6$. The next level turns out to be $i = 5, 4$. Following the branch-and-bound procedure, the decision tree shown in Figure 2.4 is produced.

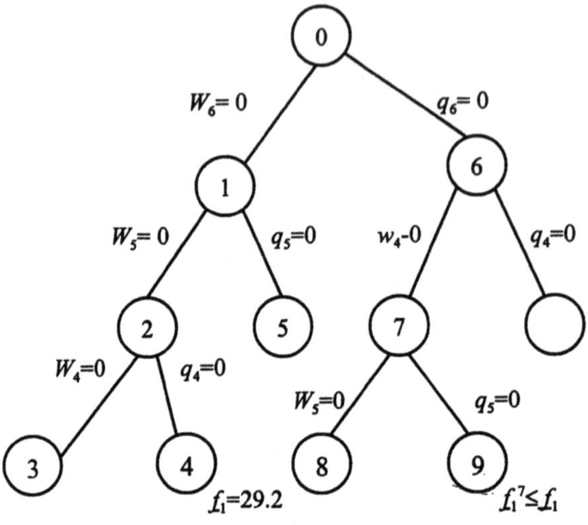

Figure 2.4. Branch-and-bound tree, Example 2.5 [Bard and Moore 1990]

The optimum results obtained are:

$X = (0, 0.9)$
$Y = (0. 0.6, 0.4)$
$W = (0, 1, 3, 6, 0, 0)$
$F_1 = 29.2$

Bard and Moore [1990] solved a wide range of problems containing up to 60 upper-level variables, 40 lower-level variables, and 40 constraints for the linear case. They also examined the influences of the various variables and factors such as the number of lower level decision variables, ill-conditioned problems, and different branching rules. These investigators concluded that the proposed branch-and-bound approach compared favorably, both in performance and in robustness, with virtually all other contenders at that time. Furthermore, due to the very general nature of the branch-and-bound approach, the basic algorithm can be extended so that it can be used to solve non-linear problems. However, Ben-Ayed [1993] stated that, for large-scale problems, the efficiency of the approach is still constrained by the exponential growth of the branch-and-bound tree.

2.3.3.1 Algorithm of Hansen et al.

Hansen *et al.* [1992] formed a new branch-and-bound algorithm by examining the tightness of the follower's constraint and based on which to form the branch-and-bound tree and to fathom or simplify the sub-problem. We shall follow Hansen *et al.* [1992] in the following discussions and the numerical example.

For the ease of discussion, consider the following linear BLPP with constraints added to the upper-level problem:

$$\max_x f_1(x, y) = c_{11}x + c_{12}y \tag{2.18}$$

subject to:

$$B_1 x + B_2 y \leq d$$
$$x \geq 0$$

where y solves:

$$\max_y f_2(x, y) = c_{21} x + c_{22} y \tag{2.19}$$

subject to:

$$A_1 x + A_2 y \leq b$$
$$y \geq 0$$

where x and y are n_1- and n_2-dimensional vectors, respectively, and B_1 and A_1 are $m_1 \times n_1$- and $m_2 \times n_2$-dimensional matrices, respectively. For simplicity, the transpose sign T for the coefficients of the objective functions have been omitted. The above model contains first-level constraints that are binding only for the leader but also influenced by the decision of the follower. In addition to the necessary theory for the algorithm, these investigators also proved that the problem represented by Equations 2.18 and 2.19 is strongly NP-hard. The authors first introduced the 0-1 Boolean variable α_i, which equals 1 if constraint i is tight or is an active constraint, and equals to 0 otherwise. The following necessary conditions for optimality was established:

Theorem 2.1. In any rational solution to the BLPP represented by Equations 2.18 and 2.19, the tightness of the constraints in the follower's sub-problem is such that:

$$\sum_{i|(A2)ij} \alpha_i \geq 1, \text{ if } (c_{22})_j > 0 \tag{2.20}$$

$$\sum_{i|(A2)_{ij}<0} \alpha_i + \alpha_{m2+j} \geq 1, \text{ if } (c_{22})_j < 0 \tag{2.21}$$

for $j = 1, 2, \ldots, n_2$. Furthermore, the investigators also proved that in any optimal solution to this BLPP problem, the tightness of the constraints in the follower's sub-problem is such that Conditions 2.20 and 2.21 are satisfied for all $j \in (1, 2, \ldots, n_2)$ according to whether $c_{22} > 0$ and $c_{22} < 0$, respectively.

If the ith constraint is tight, that is $\alpha_i = 1$, then the ith constraint of the lower becomes an equality and it can be eliminated if it is a non-negative constraint. However, this approach does not reduce the structure constraints. Obviously, in case all the variables, x_2, for the lower level are eliminated, the problem reduced to a one-level problem and the objective function of the lower level can be ignored.

If $\alpha_i = 0$, the ith constraint in the follower's problem becomes a strict inequality if it is a non-negativity constraint. From the complementary slackness theory, the ith variable of the dual of the follower's problem must be equal to zero. Thus, the dual problem will be solved.

Based on Equations 2.20 and 2.21, a branch-and-bound algorithm with depth-first search was formulated. When a branch corresponding $\alpha_i = 0$ is explored, the corresponding α_i is deleted in Equations 2.20 and 2.21. When branching is done on the values of α_i, at most $2^{m2+n2+1}$-1sub-problems will be generated. This number can be reduced when $\alpha_i = 0$.

The bounds for the branch-and-bound algorithm are obtained by solving the relaxed linear programming problem. During the iteration, deleting the objective function of the follower, we obtain the leader relaxation (LR) problem. Solving this linear programming, or solving the LR problem, we obtain the upper bound.

The branch-and-bound algorithm is essentially based on the repeated solving three linear programming problems, namely, the leader's relaxed (LR) problem, the follower's relaxed (FR) problem, and the follower's sub-problem (FS). The relaxed problems are not the original problem as listed in Equations 2.18 and 2.19 but are problems during the iteration, where some or all of the y have been eliminated and the initial constraints have been separated into tight, loose, or undecided. If the follower's objective function is deleted from this iterated BLPP, we obtain the LR problem. If the problem is consisted only of the iterated follower's problem with fixed or given x, we obtain the FR problem. The FS is consisted of original follower's problem, Equation 2.19, with fixed or given x.

A subset of logical relation R is defined as:

$$R = r_k, k \in K \qquad (2.22)$$

$$r_k \equiv \sum_{i \in I_k} \alpha_i \geq 1, \; I_k \subseteq \{1,2,\ldots,m_2 + n_2\}$$

where r_k is called a logical relation. The algorithm can now be summarized in the following [Hansen *et al.* 1992].

Step 1. Initialization. Obtain an initial solution (x_{opt}, y_{opt}) by a heuristic method. If no heuristic solution, use the best guessed values as (x_{opt}, y_{opt}) and set $z_{opt} = -\infty$.

Step 2. First direct optimality test. Solve LR problem. Let the solution be represented by (x_L*, y_L*). If $z_L^* = c_{11}\, x_L^* + c_{12}\, y_L^* \leq z_{opt}$, go to Step 13. Otherwise, go to the next step.

Step 3. First direct feasibility test. Solve the dual of FR(x) problem. If it has no feasible solution, go to Step 13. Otherwise, go to the next step.

Step 4. Direct resolution test, first part. Check if the solution of LR, (x_L*, y_L*), obtained in Step 2 is rational for the current sub-problem. Solve FR(x_L*) and let the solution be y_F*. If $c_{12}\, y_F^* = c_{12}\, y_L^*$, then ($x_L$*, y_L*) is a rational solution. Go to the next step. Otherwise, go to Step 6.

Step 5. Direct resolution test, second part. Check if the LR solution is also rational for the initial problem. Solve FS(x_L*) and let the solution be y_{FS}*. If $c_{12}\, y_{FS}^* = c_{12}\, y_L^*$, then ($x_L$*, y_L*) is a rational solution. Update (x_{opt}, y_{opt}) and z_{opt}. If $z_{opt} < c_{11}\, x_L^* + c_{12}\, y_L^*$, go to Step 13. Otherwise go to Step 6.

Step 6. Second direct optimality test. Compute all the penalties p_i for strictly positive slack variables in the iteration tableau of LR. Set p_i equal to 0 for all other variables. Then for all k such that $r_k \in R$, compute:

$$\pi_k = \min_{i \in \in} p_i$$

If $z_{opt} \geq z_L^* - \max_k \pi_k$, go to Step 13. Otherwise, go to the next step.

Step 7. Second direct feasibility test. If LR is infeasible, go to Step 13.

Step 8. First conditional optimality test. Consider again LR with the penalty p_i. For all i such that If $z_{opt} \geq z_L^* - p_i$, fix α_i at 0 and update R.

Step 9. Third direct optimality test. If R contains a relation r_k such that $\alpha_j = 0$ for $j \in I_k$, go to Step 13.

Step 10. Relational optimality test. For all remaining y_j appearing in $f_2(x, y)$, add to R the logical relations 2.20 or 2.21, if they are non-redundant. Eliminate from R those relations that have become redundant.

Step 11. Second conditional optimality test. If R contains a relation r_k such that $\alpha_j =$ 0 for all $j \in I_k$ except for one index i, set the corresponding $\alpha_i = 1$. Return to Step 2

Step 12. Branching. Apply the selected branching rule to choose either a free variable α_l, or, a logical relation r_k, for which all variables α_i with $i \in I_k$ are free. In the former case, branching by fixing $\alpha_i = 1$. In the latter case, branching by fix the first variable α_i in r_k equal to 1. Return to Step 2.

Step 13. Backtracking. If branching took place on a variable, find the last α_i, based on which branched. Set this $\alpha_i = 0$ and free the α_j fixed at 0 after α_i was fixed at 1. Otherwise, consider the last logical relation r_k for which the branches explored are less than $m_2 + n_2$. If there is no such variable or logical relation, stop. Otherwise, update the current sub-problem and return to Step 2.

To illustrate the algorithm, consider the following example, which was solved by Hansen *et al.* [1992].

Example 2.6.

$$\max_x f_1 = 8x_1 + 4x_2 - 4y_1 + 40y_2 + 4y_3$$

subject to:

$$x_1 + 2x_2 - y_3 \leq 1.3$$
$$x_1 \geq 0$$
$$x_2 \geq 0$$

$$\max_y f_2 = -2y_1 - y_2 - 2y_3$$

subject to:

$$-y_1 + y_2 + y_3 \leq 1 \qquad (\alpha_1)$$
$$4x_1 - 2y_1 + 4y_2 - y_3 \leq 2 \qquad (\alpha_2)$$

$$4x_2 + 4y_1 - 2y_2 - y_3 \leq 2 \quad (\alpha_3)$$
$$y_1 \geq 0 \quad (\alpha_4)$$
$$y_2 \geq 0 \quad (\alpha_5)$$
$$y_3 \geq 0 \quad (\alpha_6)$$

Step 1. The initial solution is obtained by using the following equation:

$$\max_{x,y} z = \lambda\, c_{11} x + (1-\lambda)\, c_{22}\, y$$

where $\lambda = (n_1 + n_2)/(n_1 + 2n_2) = 5/7$
thus, $(x_{opt}, y_{opt}) = (1.5, 0, 1, 0, 2)$ and $z_{opt} = 9$.

Step 2. Solve LR, obtain $(x_L^*, y_L^*) = (0, 0, 1.5, 1.5, 1)$ with $z_L^* = 58$ which is larger than $z_{opt} = 9$.

Step 3. Since no constraint has imposed, it is feasible.

Step 4. Solve FR, obtain $y_F^* = (0, 0, 0)$. Since $c_{12}\, y_L^* = -6.5 \neq y_F^* = 0$, This solution is not rational. Go to Step 6.

Step 6. The penalties p_i associated with the follower's constraints are:

$$p_1 = 0, p_2 = 0, p_3 = 0, p_4 = 28.8, p_5 = 42, p_6 = 22$$

Step 10. The set of R logical conditions are:

$$\alpha_1 + \alpha_2 + \alpha_3 \geq 1$$
$$\alpha_3 + \alpha_5 \geq 1$$
$$\alpha_2 + \alpha_3 + \alpha_6 \geq 1$$

Step 12. Branch on the second relation. Since $p_5 > p_3$, first consider $p_5 = 1$. This leads to $y_2 = 0$ and the problem reduced to:

$$\max_x f_1 = 8x_1 + 4x_2 - 4y_1 + 4y_3$$

subject to:

$$x_1 + 2x_2 - y_3 \leq 1.3$$
$$x_1 \geq 0$$
$$x_2 \geq 0$$

$$\max_y f_2 = -2y_1 - 2y_3$$

subject to:

$$-y_1 + y_3 \leq 1$$
$$4x_1 - 2y_1 - y_3 \leq 2$$
$$4x_2 + 4y_1 - y_3 \leq 2$$
$$y_1 \geq 0$$
$$y_3 \geq 0$$

Step 2. The solution of the LR problem is $(x_L, y_L^*) = (1.5, 0, 1, 2)$ with $z_L^* = 16 \le z_{opt}$. The incumbent solution remains $z_{opt} = 16$ and $(x_{opt}, y_{opt}) = (1.5, 0, 1, 2)$.

Step 13. Backtracking leads to the branches: $\alpha_5 = 0$, $\alpha_3 = 1$. Variable y_1 is eliminated. The sub-problem is reduced to:

$$\max_x f_1 = 8x_1 + 8x_2 + 38y_2 + 3y_3 - 2$$

subject to:

$$x_1 + 2x_2 - y_3 \le 1.3$$
$$x_1 \ge 0$$
$$x_2 \ge 0$$

$$\max_y f_2 = 2x_2 - 2y_2 - 2.5y_3 - 1$$

subject to:

$$x_2 + 0.5y_2 + 0.75y_3 \le 1.5 \quad (\alpha_1)$$
$$2x_1 + x_2 + 1.5y_2 - 0.75y_3 \le 1.5 \quad (\alpha_2)$$
$$x_2 - 0.5y_2 - 0.25y_3 \le 0.5 \quad (\alpha_4)$$
$$y_2 \ge 0 \quad (\alpha_5)$$
$$y_3 \ge 0 \quad (\alpha_6)$$

Tightness of the fourth constraint, $\alpha_4 = 1$ allows y_2 to be eliminated. The sub-problem now becomes:

$$\max_x f_1 = 8x_1 + 84x_2 - 16y_3 - 40$$

subject to:

$$x_1 + 2x_2 - y_3 \le 1.3$$
$$x_1 \ge 0$$
$$x_2 \ge 0$$

$$\max_y f_2 = -2x_2 - 1.5y_3 + 1$$

subject to:

$$2x_2 + 0.5y_3 \le 2 \quad (\alpha_1)$$
$$2x_1 + 4x_2 - 1.5y_3 \le 3 \quad (\alpha_2)$$
$$-2x_2 + 0.5y_3 \le -1 \quad (\alpha_5)$$
$$y_3 \ge 0 \quad (\alpha_6)$$

The new logical relation is:

$$\alpha_2 + \alpha_6 \ge 1$$

Solving the current LR problem, we obtain: $(x_L, y_L^*) = (0, 0.833, 0.466)$ with $z_L^* = 26.73 > z_{opt}$, and $p_1 = 0$, $p_2 = 8.333$, $p_6 = 12..333$. Solve FR yields: $y_F^* = 0.3555$ and $c_{12}\, y_F = -1.3 \neq c_{12}\, y_L = -1.466$. Thus, (x_L, y_L^*) is not rational.

According to Step 8, penalty $p_6 = 12.33$, which is such that $z_{opt} > z_L^* - p_6$. Thus, $\alpha_6 = 0$. The set R is now reduced to $\alpha_2 \geq 1$. Tightness of the second constraint allows y_3 to be eliminated. Thus, the problem now becomes:

$$\max_x f_1 = -13.33x_1 + 41.33x_2 - 8$$

subject to:

$$-x_1 - 2x_2 \leq -2.1$$
$$x_1 \geq 0$$
$$x_2 \geq 0$$

$$\max_y f_2 = -2x_2 - 6x_2 + 4$$

subject to:

$$2x_1 + 10x_2 \leq 9 \qquad (\alpha_1)$$
$$x_1 - x_2 \leq 0 \qquad (\alpha_5)$$
$$-2x_2 - 4x_2 \leq -3 \qquad (\alpha_6)$$

Solving the current LR problem, we obtain $x_L^* = (0.5, 0.8)$ with $z_L^* = 18.4$. Solve the FS, obtain $y_F^* = (0, 2, 8)$ with $c_{12}y_L^* = c_{12}y_F^* = -1.8$. Thus, the solution (0.5, 0.8, 0, 2, 0.8) is rational for the initial problem. Update the problem and backtracking discovers the current solution is the optimal solution. Thus, the problem is solved by only examining three nodes.

Hansen *et al.* [1992] carried out a fairly extensive numerical experiment. Problems with up to 150 constraints, 250 variables controlled by the leader and 150 variables controlled by the follower were solved. Experiments to compare the current approach with that of Bard and Moore [1990] and Judice and Faustino [1992] were also carried out. The current approach always outperforms than that of Bard and Moore.

2.3.4 Penalty-function Approach

Based on the fact that when the duality gap is reduced to zero, the optimal solution of the problem will be reached; Anandalingam and White [1990] proposed a penalty-function approach by appending this gap to the upper-level objective function with a penalty. This structure leads to the decomposition of the composite problem into a series of linear programs and thus an efficient algorithm can be developed. With certain assumptions, these investigators derived the necessary theories for this approach. Again, we shall not go into details of the theoretical developments. The following discussion follows the work of Anandalingam and White [1990].

Given the upper-level control vector x_1, the lower-level problem, Equation 2.2, reduced to the following one-level regular linear programming problem:

$$\max c_{22}^{T} x_2 \tag{2.23}$$

subject to:

$$A_2 x_2 \le b - A_1 x_1$$
$$x_2 \ge 0$$

which is the primal problem. The dual problem of this primal problem is:

$$\min w^{T} (b - A_1 x_1) \tag{2.24}$$

subject to:

$$w^{T} A_2 \ge c_{22}$$
$$w \ge 0$$

where w is the dual row vector. The optimal solution lies in the interval $[c_{22}^{T}x_2, w^{T}(b - A_1 x_1)]$. That is, the optimum lies between the maximum of the primal and the minimum of the dual. When the duality gap given by the difference:

$$\Pi(x_1, x_2, w) = [w^{T}(b - A_1 x_1) - c_{22}^{T}x_2]$$

is equal to zero, the lower-level optimal solution would be reached. Thus, use the above difference as a penalty term, the following overall problem can be obtained:

$$\max_{x1,x2,w} F(x_1, x_2, w, K) = c_{11}^{T} x_1 + c_{12}^{T} x_2 - K [w^{T} (b - A_1 x) - c_{22}^{T} x_2] \tag{2.25}$$

subject to:

$$A_1 x_1 + A_2 x_2 \le b$$
$$w^{T} A_2 \ge c_{22}$$
$$x_1, x_2, w \ge 0$$

where K is a large positive constant and the above expression will reach optimum when $[w^{T} (b - A_1 x_1) - c_{22}^{T} x_2]$ approaches zero.

By assuming the feasible region for both the primal vectors, x_1 and x_2, and the dual vector, w, forming nonempty bounded polyhedrons, Anandalingam and White [1990] developed an algorithm, which converges to the optimum by discretely increasing the value of the penalty K starting with K=0. To form the search procedure, these investigators proved that the following expression is convex:

$$\Theta(w, K) = max_{x1, x2} F(x_1, x_2, w, K) \tag{2.26}$$

subject to:

$$A_1 x_1 + A_2 x_2 \leq b$$
$$x_1, x_2 \geq 0$$

Since the function $\Theta(w, K)$ is convex, a solution to the problem:

$$\max_w \Theta(w, K) \tag{2.27}$$

subject to:

$$w^T A_2 \geq c_{22}$$
$$w \geq 0$$

can be obtained by searching the extreme corners of the feasible set of the dual variable w. Thus, the algorithm essentially is, at each fixed value of K, searching the vertexes of the polyhedron for the dual variable w. After the desirable value has been obtained at this fixed K, the value of K can be increased and the search is carried out again until the duality gap reaches zero. Based on the convexity of the function $\Theta(w, K)$, the optimality condition at any iteration i can be obtained from the following parametric function with K and w as parameters:

$$\chi (w^i, K) = \min_w (w - w^i)(b - A_1 x_1(w^i, K)) \tag{2.28}$$

subject to:

$$w^T A_2 \geq c_{22}$$
$$w \geq 0$$

Thus, Anandalingam and White [1990] proposed the following procedure with small discrete increases of K at each cycle:

Step 0. (Initialization) Set $i = 0$ and $K = 0$. Choose w^0 within the feasible region, and choose a small step size λ'. Go to Step 1.

Step 1. Solve the following problem with fixed values of w and K to obtain $\{x_1(w^i, K), x_2(w^i, K)\}$:

$$\max_{x1, x2} \{c_{11}{}^T x_1 + c_{12}{}^T x_2 - K [w^T (- A_1 x_1) - c_{22}{}^T x_2]\} \tag{2.29}$$

subject to:

$$A_1 x_1 + A_2 x_2 \leq b$$
$$w^T A_2 \geq c_{22}$$
$$x_1, x_2, w \geq 0$$

Notice that the maximization is with respect to x_1 and x_2 only.

Step 2. Solve Equation 2.29 to obtain $w^*(w^i, K)$.

Step 3. (Optimality Test)

(3.1) If $\chi(w^i, K) < 0$, then set $w^{i+1} = w^*(w^i, K)$ and $i = i + 1$. Go to Step 1.
(3.2) If $\chi(w^i, K) \geq 0$, and $\pi\{x_1(w^i, K), x_2(w^i, K), w^i\} > 0$, then set $K = K + \lambda'$ and I = 0. Go to Step 1.
(3.3) If $\chi(w^i, K) \geq 0$, and $\pi\{x_1(w^i, K), x_2(w^i, K), w^i\} = 0$, then optimality is reached. The optimal solution is $\{x_1(w^i, K), x_2(w^i, K), w^i\}$, and stop.

To illustrate the procedure, the same linear two-level programming problem solved by Anandalingam and White [1990] is used in the following example.

Example 2.7. In the following discussions, we follow the solution obtained by Anandalingam and White [1990].

$$\max_{x1} f_1 = x_1 + 3x_2$$

where x_2 solves:

$$\max_{x2} f_2 = x_1 - 3x_2$$

subject to:

$$\begin{aligned} -x_1 - 2x_2 &\leq -10 \\ x_1 - 2x_2 &\leq 6 \\ 2x_1 - x_2 &\leq 21 \\ x_1 + 2x_2 &\leq 38 \\ -x_1 + 2x_2 &\leq 18 \\ x_1, x_2 &\geq 0 \end{aligned}$$

For simplicity, X will be used to denote the above constraint set. The dual constraints of the lower-level problem are:

$$\begin{aligned} W &= (w_1, w_2, w_3, w_4, w_5)^{\mathrm{T}} \\ &= -2w_1 - 2w_2 - w_3 + 2w_4 + 2w_5 \geq -3 \\ w_i &\geq 0, i = 1, \ldots, 5 \end{aligned}$$

The solution procedure proceeds as follows [Anandalingam and White 1990]:

Iteration 1.

Step 0. (Initialization) Let $i = 0$, $K = 0$, $w^0 = (1.5, 0, 0, 0, 0)$ and $\lambda' = 0.1$.

Step 1. To solve Equation 2.29, we first obtain the coefficients for the variables in the objective function. For x_1, we have $c_{11} + K\, w^0 A_1 = 1$ and for x_2, we have $c_{12} + K\, c_{22} = c_{12} = 3$. Thus, Equation 2.29 becomes:

$$\max_{x1,x2}\ \{x_1 + 3\,x_2\} \text{ and subject to the constraint of } X$$

The solution is $x_1\,(w^0, K = 0) = 10$ and $x_2\,(w^0, K = 0) = 14$. Go to Step 2.

Step 2. Substituting the values obtained in Step 1 into Equation 2.28, we have:

$$\chi\,(w^i, K) = \min_w\ \{-4w_2 + w_3 + 28w_4 + 28w_5\}$$

subject to:

$$-2w_1 - 2w_2 - w_3 + 2w_4 + 2w_5 \geq -3$$
$$w_1, w_2, w_3, w_4, w_5 \geq 0$$

The solution for this problem is $(w_1, w_2, w_3, w_4, w_5) = (0, 1.5, 0, 0, 0)$.

Step 3. (Optimality Test)

$\chi\,(w^0, K = 0) = -6 < 0$, thus set $(w^{0+1})^T = (w^1)^T = (w^*, K= 0)^T = (0, 1.5, 0, 0, 0)$ and $i = i + 1 = 0 + 1 = 1$. Go to Step 1.

Iteration 2.

Step 1. The coefficients are $(x_1, x_2) = (1, 3)$. Thus, Equation 2.29 becomes:

$$\max_{x1,x2}\ \{x_1 + 3x_2\ |(x_1, x_2) \in X\}$$

The solution is $x_1(w^1, K = 0) = 10$ and $x_2(w^1, K = 0) = 14$. Go to Step 2.

Step 2. The solution remains unchanged. Go to Step 3.

Step 3. (Optimality Test)

$\chi\,(w^1, K = 0) = 0 \geq 0$, and duality gap $\pi\{x_1(w^1, K = 0), x_2(w^1, K = 0), w^1\} = 36 > 0$, then the optimal solution is not found. Set $K = 0 + 0.1 = 0.1$. Go to Step 1.

Iteration 3.

Step 1. The coefficients for x_1 and x_2 are $c_{11} + K\,w^1\,A_1 = 1.15$ and $c_{12} + K\,c_{22} = 2.7$, respectively. Thus, Equation 2.29 becomes:

$$\max_{x1,x2}\ \{1.15\,x_1 + 2.7\,x_2\,|\,(x_1, x_2) \in X\}$$

The solution is $x_1(w^1, K= 0.1)= 10$ and $x_2(w^1, K= 0.1)= 14$. Go to Step 2.

Step 2. The solution remains unchanged. Go to Step 3.

Step 3. (Optimality Test)

The result remains unchanged. Set $K = K + \lambda' = 0.1{+}0.1 = 0.2$. Go to Step 1.

Iteration 4.

Step 1. $c_{11} + K\,w^1\,A_1 = 1.3$ and $c_{12} + K\,c_{22} = 2.4$. Thus, Equation 2.29 becomes:

$$\max_{x1,x2} \{1.3\, x_1 + 2.4\, x_2 \mid (x_1, x_2) \in X\}$$

The solution is $x_1(w^1, K = 0.2) = 16$ and $x_2(w^1, K = 0.2) = 11$. Go to Step 2.

Step 2. Substituting the values obtained from Step 1 into Equation 2.28, we have:

$$\chi\,(w^i, K) = \min_w \{6w_1 - 10w_2 - 11w_3 + 22w_4 + 34w_5 + 15\}$$

subject to:

$$-2w_1 - 2w_2 - w_3 + 2w_4 + 2w_5 \geq -3$$
$$w_1, w_2, w_3, w_4, w_5 \geq 0$$

The solution for this problem is $(w_1, w_2, w_3, w_4, w_5) = (0, 0, 3, 0, 0)$. Go to Step 3.

Step 3. (Optimality Test)

$\chi\,(w^1, K = 0.2) = -18 < 0$, then set $(w^{1+1})^T = (w^2)^T = (w^*, K = 0.2)^T = (0, 0, 3, 0, 0)$ and $i = i + 1 = 1 + 1 = 2$. Go to Step 1.

Iteration 5.

Step 1. $c_{11} + K\, w^2 A_1 = 2.2$ and $c_{12} + K\, c_{22} = 2.4$. Thus, Equation 2.29 becomes

$$\max_{x1,x2} \{2.2\, x_1 + 2.4\, x_2 \mid (x_1, x_2) \in X\}$$

The solution is $x_1(w^2, K = 0.2) = 16$ and $x_2(w^2, K = 0.2) = 11$. Go to Step 2.

Step 2. The solution remains unchanged. Go to Step 3.

Step 3. (Optimality Test)

$\chi\,(w^2, K = 0.2) = 0$ and the duality gap $\pi\ \{x_1(w^2, K = 0.2), x_2(w^2, K = 0.2), w^2\} = 0$. Thus, optimality is reached. The optimal solution is $(x_1, x_2) = (16, 11)$.

Anandalingam and White [1990] also provided the computational results of 50 randomly selected problems. The proposed penalty-function algorithm marginally outperforms the grid-search algorithm [Bard 1982, 1983b] and easily outperforms the kth-best algorithm [Bialas and Karwan 1982, 1984]. Furthermore, White and Anandalingam [1993] extend the procedure to ten steps to obtain the global optimal solution for the linear case. Computational results show that the new ten-step penalty-function approach is easily outperformed the kth-best algorithm, and slightly worse than branch-and-bound algorithm [Bard and Moore 1990].

Aiyoshi and Shimizus [1981a, 1981b, 1984] proposed a penalty-function approach for solving the BLPP with non-linear objective functions. However, their penalty function is convex, which is much easier to solve than the resulting non-convex function from the KTT transformations.

2.4 Discussions

Only linear BLPP is discussed in this chapter. Linear BLDPP and TLPP together with non-linear problems will be discussed in the next chapter. Another promising area, which also will be covered in the next chapter, is the evolutionary approach.

By comparing the results of the solutions between Examples 2.1 and 2.3, an interesting point can be observed. In Example 2.1, where the kth-best algorithm is used, the results obtained are $x = (8, 3)$ and $f = (13, 14)$ which is at vertex C (see Figures 2.1 and 2.2). While when the mixed-integer approach of Fortuny-Amat and McCarl is used to solve the same problem, the optimum is at vertex B with $x = (7.5, 1.5)$ and $f = (13.5, 10.5)$. In other words, the results really depend on which player moves first. From another more rational standpoint, the results depend on which player is more important or influence more on the overall results. If we assume both players are equally important, then the optimum should lies between these two vertices. If we borrow the concept used in multiple-objective optimization, compromise between the two players should be used. This compromise is certainly not at the vertex. In other words, we are really asking too much of a fuzzy aspects and end-up with extra unnecessary computations.

Chapter 3

Other Multi-level Programming Algorithms

Although not as efficient, some of the algorithms discussed in the last chapter also can be extended or modified for solving more complicated problems such as MLDPP, MLPP, quadratic MLPP, and even non-linear BLPP. In fact, several algorithms discussed in Chapter 2 were formulated for solving quadratic or linear-quadratic problems. Furthermore, to solve non-linear and discrete problems, heuristic approaches such as descend or gradient approach, branch-and-bound technique, and cutting-plane algorithm have been developed. The use of the more recently developed evolutionary approaches, such as genetic algorithms, tabu search, and simulated annealing have also been proposed. Some of these algorithms will be discussed in this chapter.

3.1 Linear Bi-level Distributed Programming

A linear bi-level distributed programming problem (BLDPP) is characterized by one decision center at the top level and p divisions or decision centers at the bottom level with the assumption that decision units in the lower level are independent and under the control of the upper level. In other words, the leader or top level makes the decision first and the followers must make their decisions based on the leader's decision. Depending on the actual system, different degrees of cooperation between the followers can happen. In an organization, the divisions or followers are required to cooperate for the common good in addition to optimize individually. On the other hand, many duo-ploy systems such the traffic system, there is no cooperation between the various traffic users, who are the followers, except under special situations. The same is true for most economic duo-ploy systems.

Let $f_{ki}(x)$ represents the objective function of the ith division at the kth level and c_{kij} represents the cost coefficient of the decision variable x_j for the ith division of the kth level, with $k = 1,2$ and $i = 1, 2,\ldots,s_k$.Since there is only one division at the top level, thus $s_{k=1} = 1$. Assume p divisions at level 2, then $i = 1, 2, ..., s_k$ and $s_2 = p$. The BLDPP problem can be represented as:

$$\max_{x11} f_{11}(x) = \Sigma_j c_{11j}^T x_j \qquad \text{(upper level)} \qquad (3.1)$$

where $x_{21}, x_{22}, ..., x_{2p}$ solve:

$$\begin{cases} \max_{x21} f_{21}(x) = \Sigma_j c_{21j}^T x_j \\ \ldots \\ \max_{x2p} f_{2p}(x) = \Sigma_j c_{2pj}^T x_j \end{cases} \qquad \text{(lower level)}$$

subject to:

$$\Sigma_{k,i} A_{ki} x_{ki} \le b, \ k = 1, 2 \text{ and } i = 1, 2, ..., s_k$$
$$x_j \ge 0, j = 1, 2, ..., n$$

where $j = 1, 2, ..., n$ represents the jth decision variable and x_j for $j = 2, 3, ..., n$ are decisions of the p divisions in the lower level. Notice that $n - 1 = p$.

In theory, most of the algorithms discussed in Chapter 2 can be extended or modified for solving the linear BLDPP. However, in actual practice, the algorithm can become complex and difficult to solve. To illustrate the approaches, two algorithms for solving the linear BLDPP are discussed in the following.

Anandalingam [1988] proposed a procedure for solving the BLDPP by using the KKT optimality conditions and the mixed-integer approach of Fortuny-Amat and McCarl [1981]. Another algorithm for solving the multiple agents system was developed by Anandalingam and Apprey [1991] based on a penalty-function approach, which was developed for BLPP by Anandalingam and White [1990]. Anandalingam and Apprey also discussed the existence of Nash equilibrium under different conditions and methods for coordinating conflicting agents. The approach was used to solve a conflict resolution problem of an international river between India and Bangladesh.

3.1.1 Mixed-integer Problem with Complementary Slackness

Using the KKT conditions, the lower-level problems can be transformed into the constraints of the upper-level problem. The transformed problem is a one-level optimization problem. Thus, the original nested optimization problem represented by Equation 3.1 is reduced to the following traditional non-linear programming problem:

$$\max_{x11} f_{11}(x) = \Sigma_j c_{11j}^T x_j \qquad (3.2)$$

subject to:

$$\sum_{k,i} \boldsymbol{A}_{ki}\, \boldsymbol{x}_{ki} \leq \boldsymbol{b},\ k = 1, 2 \text{ and } i = 1, 2, \ldots, s_k$$
$$\boldsymbol{w}^T \left(\sum_{k,i} \boldsymbol{A}_{ki}\, \boldsymbol{x}_{ki} - \boldsymbol{b}\right) = \boldsymbol{0},\ k = 1, 2 \text{ and } i = 1, 2, \ldots, s_k$$
$$\sum_j \boldsymbol{w}\, \boldsymbol{A}_{ki} \geq \boldsymbol{c}_{2jj},\ j = 1, 2, \ldots, p$$
$$\boldsymbol{w}, \boldsymbol{x}_j \geq \boldsymbol{0},\ j = 1, 2, \ldots, n$$

where $j = 1, 2, \ldots, n$ represents the jth decision variable and $\boldsymbol{x}_j$ for $j = 2, 3, \ldots, n$ are decisions of the p divisions in the lower level. Note that $n - 1 = p$. and $\mathbf{w}$ is a dual vector.

The non-linear problem was first reduced to mixed-integer problem by using the approach due to Fortuny-Amat and McCarl [1981] and was discussed in the previous chapter. Introducing the binary variable, $\eta \in \{0, 1\}$ and let $\boldsymbol{q}$ represent the original constraint matrix, the non-linear complementary slackness constraint:

$$w^T q = w^T \left(\sum \boldsymbol{A}_{ki}\, \boldsymbol{x}_{ki} - \boldsymbol{b}\right) = \boldsymbol{0}$$

is replaced by the following two inequalities:

$$q \leq M\eta$$
$$w \leq (1\text{-}\eta)M$$

where M is a large positive constant. Using the above inequalities to enforce the complementary slackness conditions, a much larger mixed-integer problem is formed. Let the second constraint set in Equation 3.2:

$$\boldsymbol{w}^T \left(\sum_{k,i} \boldsymbol{A}_{ki}\, \boldsymbol{x}_{ki} - \boldsymbol{b}\right) = \boldsymbol{0},\ \ k = 1, 2 \text{ and } i = 1, 2, \ldots, s_k$$

be replaced by the two inequalities:

$$\boldsymbol{w} \leq (1\text{-}\eta)\, M \text{ and}$$
$$\sum_{k,i} \boldsymbol{A}_{ki}\, \boldsymbol{x}_{ki} - \boldsymbol{b} \leq M\, \eta,\ \ k = 1, 2 \text{ and } i = 1, 2, \ldots, s_k$$

where η is a binary vector and Equation 3.2 becomes:

$$\max_{x11} f_{11}(\boldsymbol{x}) = \sum_j \boldsymbol{c}_{11j}{}^T \boldsymbol{x}_j \tag{3.3}$$

subject to:

$$\sum_{k,i} \boldsymbol{A}_{ki}\, \boldsymbol{x}_{ki} \leq \boldsymbol{b},\ k = 1, 2 \text{ and } i = 1, 2, \ldots, s_k$$
$$\boldsymbol{w} \leq (1\text{-}\eta)M$$
$$\sum_{k,i} \boldsymbol{A}_{ki}\, \boldsymbol{x}_{ki} - \boldsymbol{b} \leq M\, \eta$$
$$\sum_j \boldsymbol{w}\, \boldsymbol{A}_{ki} \geq \mathbf{c}_{2jj},\ \ j = 1, 2, \ldots, p$$
$$\eta \in \{0, 1\}$$
$$\boldsymbol{w}, \boldsymbol{x}_j \geq \boldsymbol{0},\ \ j = 1, 2, \ldots, n$$

where M is a large positive constant, $j = 1, 2, \ldots, n$ represents the jth decision variable, and $\boldsymbol{x}_j$ for $j = 2, 3, \ldots, n$ are decisions of the p divisions in the lower level.

Equation 3.3 represents a mixed-integer problem. The following example illustrates the procedure.

Example 3.1. A problem with two lower levels considered by Anandalingam [1988] is solved in the following.

$$\max_{x1} f_1 = x_1 + y_1 + 2\,y_2 + y_3$$

where y_1, y_2, and y_3 solve:

$$\begin{cases} \max_{y1} f_{21} = x_1 + 3\,y_1 + y_2 + y_3 \\ \max_{y2} f_{22} = x_1 + y_1 + 3\,y_2 + y_3 \\ \max_{y3} f_{23} = x_1 + y_1 + y_2 + 3\,y_3 \end{cases}$$

subject to:

$$\begin{aligned} &3x_1 + 3y_1 \le 30 \\ &2x_1 + y_1 \le 20 \\ &y_1 + y_3 \le 15 \\ &x_1 + 2y_1 + 2y_2 + y_3 \le 40 \\ &y_1 \le 10 \\ &y_3 \le 10 \\ &x_1, y_1, y_2, y_3 \ge 0 \end{aligned}$$

Applying the KKT conditions, we have:

$$\max f_1 = x_1 + y_1 + 2y_2 + y_3$$

subject to:

$$\begin{aligned} &w_1\,(3x_1 + 3y_1 - 30) = 0 \\ &w_2\,(2x_1 + y_1 - 20) = 0 \\ &w_3\,(y_1 - 10) = 0 \\ &w_4\,(y_1 + y_3 - 15) = 0 \\ &w_5\,(y_3 - 10) = 0 \\ &w_6\,(x_1 + 2y_1 + 2y_2 + y_3 - 40) = 0 \\ &3w_1 + w_2 + 2w_6 = 3 \\ &w_3 + w_4 + 2w_6 = 3 \\ &w_4 + w_5 + w_6 = 3 \\ &w_1, w_2, w_3, w_4, w_5, w_6 \ge 0 \\ &x_1, y_1, y_2, y_3 \in \boldsymbol{Y} \end{aligned}$$

where $\boldsymbol{Y}$ represents the constraint set. Applying the mixed-integer approach of Fortuny-Amat and McCarl [1981], we finally obtain the following expression:

$$\max f_1 = x_1 + y_1 + 2y_2 + y_3$$

subject to:

$$
\begin{aligned}
& w_1 \leq (1-\eta_1) \\
& 3x_1 + 3y_1 - 30 \leq M\eta_1 \\
& w_2 \leq (1-\eta_2) \\
& 2x_1 + y_1 - 20 \leq M\eta_2 \\
& w_3 \leq (1-\eta_3) \\
& (y_1 - 10) \leq M\eta_3 \\
& w_4 \leq (1-\eta_4) \\
& (y_1 + y_3 - 15) \leq M\eta_4 \\
& w_5 \leq (1-\eta_5) \\
& (y_3 - 10) \leq M\eta_5 \\
& w_6 \leq (1-\eta_6) \\
& (x_1 + 2y_1 + 2y_2 + y_3 - 40) \leq M\eta_6 \\
& 3w_1 + w_2 + 2w_6 = 3 \\
& w_3 + w_4 + 2w_6 = 3 \\
& w_4 + w_5 + w_6 = 3 \\
& \eta_1, \eta_2, \eta_3, \eta_4, \eta_5, \eta_6 \in \{0, 1\} \\
& w_1, w_2, w_3, w_4, w_5, w_6 \geq 0 \\
& x_1, y_1, y_2, y_3 \in \boldsymbol{Y}
\end{aligned}
$$

where M is a large positive number. Using the LINGO programming code [1992], the above mixed-integer problem was solved. With M = 10,000, the solution obtained is $f_{11}^* = 35$, $(x_1^*, y_1^*, y_2^*, y_3^*) = (5, 5, 10, 5)$, $f_{21}^* = -5$, $f_{22}^* = 15$, and $f_{23}^* = -5$. The problem was also solved with M = 100,000, 1,000,000, 10,000,000, and 100,000,000, the answers for the solution remain the same.

3.1.2 Penalty-function Approach

Anandalingdam and Apprey [1991] extended the penalty-function approach of Anandalingdam and White [1990] to solve the bi-level multi-agents problem. The basic idea is to add every objective of lower-level agent to the objective of upper level so that the problem becomes:

$$\max_{x,\, w} \boldsymbol{Z}(\boldsymbol{x}, \boldsymbol{w}, K) = \sum \boldsymbol{c}_{11j}^T \boldsymbol{x}_j - K \{\textstyle\sum_l [\boldsymbol{w}_l (\boldsymbol{b}-\boldsymbol{A}_{2i-} \boldsymbol{x}_{2i-}) - \sum_j \boldsymbol{c}_{2lj}^T \boldsymbol{x}_j]\} \tag{3.4}$$

subject to:

$$
\begin{aligned}
& \textstyle\sum_{k,i} \boldsymbol{A}_{ki} \boldsymbol{x}_{ki} \leq \boldsymbol{b},\ k = 1, 2 \text{ and } i = 1, 2, \ldots, s_k \\
& \boldsymbol{w}_l^T \boldsymbol{A}_l \geq \boldsymbol{c}_{2ll},\ l = 1, 2, \ldots, p \\
& \boldsymbol{w}_l \geq \boldsymbol{0},\ l = 1, 2, \ldots, p \\
& \boldsymbol{x}_j \geq \boldsymbol{0},\ j = 1, 2, \ldots, n
\end{aligned}
$$

where K is a large positive constant, $j = 1, 2, ..., n$ represents the jth decision variable, and x_j for $j = 2, 3,..., n$ are decisions of the p divisions in the lower level. Notice again that n-1=p.

Let the vectors X and W_i represent the primal feasible region and the dual feasible region, respectively. The dual feasible space $W = W_1 \times W_1 \times ... \times W_1$ is assumed to be non-empty and forms a bounded polyhedral, which is further assumed to be non-degenerate at each vertex. Then, for a given value of w and a fixed K, we can define:

$$\Theta(w, K) = \max_x Z(x, w, K) \tag{3.5}$$

subject to:

$$x \in X$$

Then, $\Theta(w, K)$ is convex in its decision space, and a solution to the problem:

$$\min_w \Theta(w, K) \tag{3.6}$$

subject to:

$$w \in W$$

will occur at some $w^* \in$ the extreme points of W.

Increase K in discrete steps of λ', Anandalingam and Apprey [1991] formed the following algorithm for solving the BLDPP, which is similar to that of Anandalingam and White [1990] discussed in the previous chapter.

Step 0. Choose fairly large value of K and $w^0 \in$ the extreme points of W. Go to Step 1.

Step 1. Solve Equation 3.5 to obtain $x(w^{0*}, K)$. Go to Step 2.

Step 2. Solve Equation 3.6 to obtain w^{0*}. If $w^{0*} = w^0$, go to Step 3. Otherwise, set $w^0 = w^{0*}$ and go to Step 1.

Step 3. (Optimality Test)
Compute the duality gap:

$$\pi\{x(w^{0*}, K), w^0\Sigma_1 [w_1^{0*} (b - A_{2i-} x_{2i-}) - \Sigma_j c_{2lj}^T x_j]\}.$$

If $\pi\{x(w^{0*}, K), w^{0*}\} = 0$, then optimality is reached. The optimum solution is $\{x(w^{0*}, K), w^{0*}\}$. Otherwise, set $K = K + \lambda'$ and go to Step 1.

The BLDPP can be solved, even though is more difficulty, in a similar manner as that illustrated in the previous chapter.

3.2 Linear Three-level Programming Problem

Because of the complexity of MLPPs, only three level problems will be discussed in this section. In fact, even with the increase of only one level, the solution of the problem is much more complicated compared to the BLPP. Thus, not much research has been carried out. One fruitful approach appears to be the interactive fuzzy decision making to be discussed in later Chapters.

The three level programming problem (TLPP), whether from the standpoint of the three-ploy Stackelberg behavior or from the interactive organizational behavior, is a very practical problem and encountered frequently in actual practice. For example, Cassidy *et al.* [1971] considered the distribution of federal budget. Federal government, the first level, allocates the money to the various states; the states, as the second level of managers, allocate money to the various cities and the city can further allocate to different projects. The behavior of the lower hierarchies are not within the control of the higher hierarchies except through legislation and existing laws, and thus the allocation of budget of the federal government has the some of the characteristics of the Stackelburg behavior.

The three level programming problem (TLPP) can be defined as a three-person, non-zero sum game with perfect information in which players move sequentially from top to bottom level. The TLPP can be formulated as:

$$\max_{x1} f_1(\boldsymbol{x}) = \mathbf{c}_{11}{}^T \boldsymbol{x}_1 + \boldsymbol{c}_{12}{}^T \boldsymbol{x}_2 + \boldsymbol{c}_{13}{}^T \boldsymbol{x}_3 \qquad (1^{st}\text{ level}) \qquad (3.7)$$

where $\boldsymbol{x}_2$ solves:

$$\max_{x2} f_2(\boldsymbol{x}) = \boldsymbol{c}_{21}{}^T \boldsymbol{x}_1 + \boldsymbol{c}_{22}{}^T \boldsymbol{x}_2 + \boldsymbol{c}_{23}{}^T \boldsymbol{x}_3 \qquad (2^{nd}\text{ level})$$

where $\boldsymbol{x}_3$ solves:

$$\max_{x3} f_3(\boldsymbol{x}) = \boldsymbol{c}_{31}{}^T \boldsymbol{x}_1 + \boldsymbol{c}_{32}{}^T \boldsymbol{x}_2 + \boldsymbol{c}_{33}{}^T \boldsymbol{x}_3 \qquad (3^{rd}\text{ level})$$

subject to:

$$(\boldsymbol{x}_1, \boldsymbol{x}_2, \boldsymbol{x}_3) \in \boldsymbol{X} = \{(\boldsymbol{x}_1, \boldsymbol{x}_2, \boldsymbol{x}_3) | \boldsymbol{A}_1 \boldsymbol{x}_1 + \boldsymbol{A}_2 \boldsymbol{x}_2 + \boldsymbol{A}_3 \boldsymbol{x}_3 \le \boldsymbol{b}, \text{ and } \boldsymbol{x}_1, \boldsymbol{x}_2, \boldsymbol{x}_3 \ge \boldsymbol{0}\}$$

where $\boldsymbol{x}_1$, $\boldsymbol{x}_2$, and $\boldsymbol{x}_3$ are the controls or decisions for the 1st, 2nd, and 3rd levels, respectively.

Equation 3.7 represents a three-level hierarchy process, whose objective is to maximize $\boldsymbol{f} = (f_1, f_2, f_3)$ by choosing the decisions $\boldsymbol{x} = (\boldsymbol{x}_1, \boldsymbol{x}_2, \boldsymbol{x}_3)$ in a top to bottom manner. In other words, the top-level player first attempts to maximize f_1 over the convex region $\boldsymbol{X}$ by selecting $\boldsymbol{x}_1$, where $\boldsymbol{X}$ represents the set of the original constraints. The second level player then maximizes f_2 based on the already decided $\boldsymbol{x}_1$ by selecting $\boldsymbol{x}_2$. The last level or third player maximizes f_3 based on the decisions of $\boldsymbol{x}_1$ and $\boldsymbol{x}_2$ by selecting $\boldsymbol{x}_3$. Thus, in a sense, the problem of the third or the lowest level player is the simplest because once $\boldsymbol{x}_1$ and $\boldsymbol{x}_2$ are decided, the problem for the third level is a simple one level programming problem. Consider the third level problem with $\boldsymbol{x}_1$ and $\boldsymbol{x}_2$ fixed [Bard 1984b]:

$$X_3^* (\boldsymbol{x}_1, \boldsymbol{x}_2) = \{\boldsymbol{x}_3 \mid \boldsymbol{x}_3 \text{ solves:}$$

$$\max (\boldsymbol{c}_{33}^T \boldsymbol{x}_3 \mid \boldsymbol{x}_3 \in S^3)\} \tag{3.8}$$

where

$$S^3 = \{\boldsymbol{x}_3 \mid \boldsymbol{A}_3 \boldsymbol{x}_3 \le \boldsymbol{b} - \boldsymbol{A}_1 \boldsymbol{x}_1 - \boldsymbol{A}_2 \boldsymbol{x}_2, (\boldsymbol{x}_1, \boldsymbol{x}_2, \boldsymbol{x}_3) \ge \boldsymbol{0}\} \tag{3.9}$$

where S^3 is called the level three inducible region and X_3^* is the level three rational reaction set in reaction to $\boldsymbol{x}_1$ and $\boldsymbol{x}_2$. After the third level, the second level player's problem with fixed with $\boldsymbol{x}_1$ can be formulated as:

$$X_2^* (\boldsymbol{x}_1) = \{(\boldsymbol{x}_2, \boldsymbol{x}_3) \mid (\boldsymbol{x}_2, \boldsymbol{x}_3) \text{ solves:}$$

$$\max (\boldsymbol{c}_{22}^T \boldsymbol{x}_2 + \boldsymbol{c}_{23}^T \boldsymbol{x}_3 \mid (\boldsymbol{x}_2, \boldsymbol{x}_3) \in S^2)\} \tag{3.10}$$

where

$$S^2 = \{(\boldsymbol{x}_2, \boldsymbol{x}_3) \mid \boldsymbol{A}_2 \boldsymbol{x}_2 + \boldsymbol{A}_3 \boldsymbol{x}_3 \le \boldsymbol{b} - \boldsymbol{A}_1 \boldsymbol{x}_1, \boldsymbol{x}_1 \ge 0, \boldsymbol{x}_2 \ge 0, \boldsymbol{x}_3 \in X_3^*\} \tag{3.11}$$

where S^2 is called the level two inducible region and $\mathbf{X}_2^*$ is the level two rational reaction set. Finally, the top level or level one player's problem can be formulated as:

$$\max (\boldsymbol{c}_{11}^T \boldsymbol{x}_1 + \boldsymbol{c}_{12}^T \boldsymbol{x}_2 + \boldsymbol{c}_{13}^T \boldsymbol{x}_3 \mid (\boldsymbol{x}_1, \boldsymbol{x}_2, \boldsymbol{x}_3) \in S^1) \tag{3.12}$$

where

$$S^1 = \{(\boldsymbol{x}_1, \boldsymbol{x}_2, \boldsymbol{x}_3) \in X \mid \boldsymbol{x}_1 \ge 0, (\boldsymbol{x}_2, \boldsymbol{x}_3) \in X_2^*\} \tag{3.13}$$

where S^1 is the level one inducible region. Thus, strictly speaking, the decision variables for the second and the third levels should be expressed as $x_2(x_1)$ and $x_3(x_2(x_1))$, respectively. Notice that the upper-level DMs may restrict the options available to the lower-level DMs and the lower MDs' decisions, although cannot directly influence the decisions of the upper levels, do eventually influence the overall and each individual DM's payoffs.

From Equations 3.8-3.13, we can see the basic problems connected with the solution of the TLPP, which is very similar to the BLPP except more complicated. For example, if either X_2^* or X_3^* or both are not singletons, multiple solutions may exist at the lower levels in the hierarchy. Thus, the higher levels may not obtain the best returns if the multi-ploy decision sequence is carried out.

Bard [1984b], by the combined use of the cutting-plane method and vertex search, proposed an algorithm for solving linear TLPPs. By examining a limited number of test problems, Bard concluded that the amount of work required finding a solution seems to grow with the square of the row dimension of the underlying polyhedral set. Wen and Bialas [1986] proposed a hybrid algorithm for solving the linear TLPP. This algorithm first adopts the kth-best algorithm, which was developed for solving BLPPs, to generate the kth-best extreme point by optimizing the first-level problem, and then using a complementary-pivot algorithm to check for feasibility of the second-level problem. The procedure may pose computational problem when the number of constraints is large. Thus, Anandalingam [1988] proposed a KKT Stackelberg-equilibrium approach to reduce the computational time. In addition, Anandalingam and Apprey (1991) provide an algorithm that combines the kth-best with the penalty-function approach for solving linear TLPPs.

Wang *et al.* [1994] suggested a method for generating non-dominated solutions for a three level decentralized system with multiple-objective headquarters. The approach is based on a revised, two-level simplex algorithm with the search of a domination tree.

3.2.1 Hybrid Extreme-point Search Algorithm

The hybrid k^{th}-best algorithm [Wen and Bialas 1986] solves the three-level programming problem by first generate the kth-best extreme point, which is obtained by maximizing f_1 over $\boldsymbol{X}$. Then a complementary-pivot algorithm is used to check the feasibility. Assume that the entire feasible region is bounded and has a finite number of extreme points, the solution procedure can be illustrated as follows [Wen and Bialas 1986]:

Step 0. Set $k = 1$and $\alpha' = 0$. Go to Step 1.

Step 1. Use the kth-best algorithm to find the kth-best solution of the first-level problem, which is:

$$\max f_1(\boldsymbol{x}) = \boldsymbol{c}_{11}{}^T \boldsymbol{x}_1 + \boldsymbol{c}_{12}{}^T \boldsymbol{x}_2 + \boldsymbol{c}_{13}{}^T \boldsymbol{x}_3$$

subject to:

$$\boldsymbol{x} \in \boldsymbol{X}$$

Let the solution obtained be represented as: $\underline{\underline{\boldsymbol{x}}}_{[K]} = (\underline{\underline{\boldsymbol{x}}}_1, \underline{\underline{\boldsymbol{x}}}_2, \underline{\underline{\boldsymbol{x}}}_3)$ and go to Step 2.

Step 2. ($\boldsymbol{S}^2$ check) With the fixed values of $\underline{\underline{\boldsymbol{x}}}_2$ and $\underline{\underline{\boldsymbol{x}}}_3$, solving the following problem:

$$\max f_3(\boldsymbol{x}) = \boldsymbol{c}_{31}{}^T \boldsymbol{x}_1 + \boldsymbol{c}_{32}{}^T \underline{\underline{\boldsymbol{x}}}_2 + \boldsymbol{c}_{33}{}^T \underline{\underline{\boldsymbol{x}}}_3$$

subject to:

$$x \in X \cap \{x \in (x_1, \underline{x}_2, \underline{x}_3)\}$$

Let the solution obtained be represented as: $(\underline{x}_1, \underline{x}_2, \underline{x}_3)$, where $\underline{x}_1$ is the newly obtained value. If $\underline{x}_1 \neq \underline{x}_1$, then $\underline{x}_{[K]} \notin S^2$. Thus, $\underline{x}_{[K]}$ is not in S^3. Set $K = K+1$and go to Step 1. If $\underline{x}_1 = \underline{x}_1$, then $\underline{x}_{[K]} \in S^2$. Go to Step 3.

Step 3. (First best Check) With fix value of $\underline{x}_3$, solving the following problem:

$$\max f_2(x) = c_{21}{}^T x_1 + c_{22}{}^T x_2 + c_{23}{}^T \underline{x}_3 \tag{3.14}$$

subject to:

$$x \in X \cap \{x \in (x_1, x_2, \underline{x}_3)\}$$

Again, let the just obtained solution be represented by: $(x_1^*, x_2^*, \underline{x}_3)$. Check if $(x_1^*, x_2^*, \underline{x}_3)$ is in S^2 by solving the following problem:

$$\max f_3(x) = c_{31}{}^T x_1 + c_{32}{}^T x_2^* + c_{33}{}^T \underline{x}_3$$

subject to:

$$x \in X \cap \{x \in (x_1, x_2^*, \underline{x}_3)\}$$

Obtain the optimal solution $(\underline{x}_1, x_2^*, \underline{x}_3)$. If $\underline{x}_1 = x_1^*$, then the first-best solution of Equation 3.14 is in S^2. Go to Step 4. If $\underline{x}_1 \neq x_1^*$, then the first-best solution $\underline{x}_{[K]}$ of Equation 3.14 is not in S^2. Set:

$$\alpha = c_{21}{}^T \underline{x}_1 + c_{22}{}^T \underline{x}_2 + c_{23}{}^T \underline{x}_3,$$

go to Step 5.

Step 4. (S^3-Check) If $(x_1^*, x_2^*) = (\underline{x}_1, \underline{x}_2)$, then $\underline{x}_{[K]} \in S^3$. Thus, $\underline{x}_{[K]}$ is the optimal solution of the three-level system, and stop. Otherwise, $\underline{x}_{[K]} \notin S^3$. Set $K = K+1$ and go to Step 1.

Step 5. (CP-Check) Solve the following problem by using the restricted basic entry procedure in the parametric complementary-pivot algorithm due to Bialas and Karwan [1984]:

$$\begin{aligned} & A_1 x_1 + A_2 x_2 + u = b - A_3 x_3 \\ & -\varepsilon H x_1 + w^T A_2 - v = c_{22} \\ & c_{21}{}^T x_1 + c_{22}{}^T x_2 \geq \alpha' + \delta \\ & u\, w = 0 \\ & x_1{}^T v = 0 \\ & x_1, x_2, u, v, w \geq 0 \end{aligned} \tag{3.15}$$

where the matrix H may be any negative definite matrix, and ε is a suitably small positive scalar, and δ is a sufficiently small positive number. If we cannot find any feasible solution for Equation 3.15, then $(\underline{x}_2, \underline{x}_3)$ is the optimal solution for the lower two-level problem for fixed

$\underline{\underline{x}}_1$, hence $\underline{\underline{x}}_{[K]} \in \boldsymbol{S}^3$ and $\underline{\underline{x}}_{[K]}$ is the optimal solution for the three-level system and stop. Otherwise, $\underline{\underline{x}}_{[K]} \notin \boldsymbol{S}^3$; set $K = K + 1$ and go to Step 1.

Let us illustrate the algorithm using the following example from Wen and Bialas [1986]:

Example 3.2.

$$\max_{x3} f_1 = -4x_1 + 3x_2 + 7x_3 \qquad (1^{st} \text{ level})$$

where x_1, x_2 solve:

$$\max_{x2} f_2 = x_2 \qquad (2^{nd} \text{ level})$$

where x_1 solve:

$$\max_{x1} f_3 = x_1 \qquad (3^{rd} \text{ level})$$

subject to:

$$\begin{aligned} &x_1 + x_2 + x_3 \le 3 \\ &x_1 + x_2 - x_3 \ge 1 \\ &x_1 - x_2 + x_3 \ge 1 \\ &-x_1 + x_2 + x_3 \le 1 \\ &x_3 \le 0.5 \\ &x_1, x_2, x_3 \ge 0 \end{aligned}$$

The first step of the hybrid algorithm is to solve the following linear program:

$$\max f_1 = -4x_1 + 3x_2 + 7x_3$$

subject to:

$$\begin{aligned} &x_1 + x_2 + x_3 \le 3 \\ &x_1 + x_2 - x_3 \ge 1 \\ &x_1 - x_2 + x_3 \ge 1 \\ &-x_1 + x_2 + x_3 \le 1 \\ &x_3 \le 0.5 \\ &x_1, x_2, x_3 \ge 0 \end{aligned}$$

The first-best solution obtained is $\underline{\underline{x}}_{[1]} = (0, 1, 0)$. Using $\boldsymbol{S}^2$-Check to determine if $\underline{\underline{x}}_{[1]}$ is an element of $\boldsymbol{S}^2$ which solves:

$$\max f_3 = x_1$$

subject to:

$$
\begin{aligned}
& x_1 + x_2 + x_3 \le 3 \\
& x_1 + x_2 - x_3 \ge 1 \\
& x_1 - x_2 + x_3 \ge 1 \\
& -x_1 + x_2 + x_3 \le 1 \\
& x_2 = 1 \\
& x_3 = 0 \\
& x_1 \ge 0
\end{aligned}
$$

In this case, the resulting solution $\underline{x} = (2, 1, 0) \ne \underline{x}_{[1]}$. Hence, $\underline{x}_{[1]} \notin S^2$. Return to Step 1 to find the next best solution.

Among the adjacent extreme points of $\underline{x}_{[1]}$, which are: {(1, 0, 0), (1, 2, 0), (0.75, 0.75, 0.5)}, the second best level-three solution is $\underline{x}_{[2]} = (0.75, 0.75, 0.5)$. Again the algorithm goes to Step 2 and rejects $\underline{x}_{[2]}$ according to $\underline{x}_{[2]} \notin S^2$.

Among the candidates, {(1, 0, 0), (1, 2, 0), (1.25, 1.25, 0.5), (1, 0.5, 0.5)}, the third best solution of the level-three problem is (1.25, 1.25, 0.5). In Step 2, the result is $\underline{x}_{[3]} = \underline{x}$, hence $\underline{x}_{[3]} \in S^2$. In Step 3 the first-best check, the algorithm finds the optimal solution of the level-two problem by solving the following problem with fixed $x_3 = 0.5$:

$$\max f_2 = x_2$$

subject to:

$$
\begin{aligned}
& x_1 + x_2 + x_3 \le 3 \\
& x_1 + x_2 - x_3 \ge 1 \\
& x_1 - x_2 + x_3 \ge 1 \\
& -x_1 + x_2 + x_3 \le 1 \\
& x_3 = 0.5 \\
& x_1, x_2 \ge 0
\end{aligned}
$$

with the resulting solution $x^* = (1.25, 1.25, 0.5)$, which also belongs to S^2 by solving the following problem and has $\underline{x}_1 = x_1^* = 1.25$:

$$\max f_3 = x_3$$

subject to:

$$
\begin{aligned}
& x_1 + x_2 + x_3 \le 3 \\
& x_1 + x_2 - x_3 \ge 1 \\
& x_1 - x_2 + x_3 \ge 1 \\
& -x_1 + x_2 + x_3 \le 1 \\
& x_2 = 1.25 \\
& x_3 = 0.5 \\
& x_1 \ge 0
\end{aligned}
$$

In Step 4 check shows that $\underline{\underline{x}}_{[3]} = x^*$, hence $\underline{\underline{x}}_{[3]} \in S^3$. Thus, (1.25, 1.25, 0.5) is a global solution.

These investigators also carried out numerical experiments. They discovered that the problem is easier to solve as the percentage of variables controlled by the third level DM is increased.

3.2.2 Mixed-integer Problem with Complementary Slackness

To solve the three-level hierarchical system represented by Equation 3.7, first consider the lower two-level problem, which is composed by the lowest level and the second level. In other words, we first ignore the highest or first level of the original problem. With a given x_1, the bottom two levels can be represented as:

$$\max_{x2} f_2\,(x_1, x_2, x_3) = c_{21}{}^T x_1 + c_{22}{}^T x_2 + c_{23}{}^T x_3 \tag{3.16}$$

where x_3 solves:

$$max_{x3} f_3\,(x_1, x_2, x_3) = c_{31}{}^T x_1 + c_{32}{}^T x_2 + c_{33}{}^T x_3$$

subject to:

$$\begin{aligned} &A_2\, x_2 + A_3\, x_3 \le b - A_1\, x_1 \\ &x_2, x_3 \ge 0 \end{aligned}$$

Applying the KKT conditions to Equation 3.16, we have:

$$\max_{x2,x3} f_2\,(\, x_1, x_2, x_3) = \mathbf{c}_{22}{}^T x_1 + \mathbf{c}_{22}{}^T x_2 + c_{23}{}^T x_3 \tag{3.17}$$

subject to:

$$\begin{aligned} &A_2\, x_2 + A_3\, x_3 \le b - A_1\, x_1 \\ &u^T A_3 = c_{33} \\ &u^T (A_2\, x_2 + A_3\, x_3 + A_1\, x_1 - b) = 0 \\ &x_2, x_3, u \ge 0 \end{aligned}$$

where $\mathbf{x}_2$ and x_3 are the controls or decisions for the 2nd- and 3rd-level DMs, respectively, and u is a dual vector. Now, consider the problem composed of the original first level and the above one-level problem. Applying again the KKT conditions to this problem, we finally obtain the following one-level transformed problem of the original three hierarchical system:

$$\max f_1\,(x_1, x_2, x_3) = c_{31}{}^T x_1 + c_{22}{}^T x_2 + c_{23}{}^T x_3 \tag{3.18}$$

subject to:

$$\begin{aligned}
&A_1 x_1 + A_2 x_2 + A_3 x_3 \le b \\
&u^T A_3 = c_{33} \\
&w^T A_2 = c_{22} \\
&w^T A_3 = c_{23} \\
&w - v - \lambda u = 0 \\
&v^T (A_1 x_1 + A_2 x_2 + A_3 x_3 - b) = 0 \\
&u^T (\mathrm{A}_1 x_1 + A_2 x_2 + A_3 x_3 - b) = 0 \\
&x_1, x_2, u, w, v, \lambda \ge 0
\end{aligned}$$

where v and λ are Lagrange multipliers.

Although we only illustrated the transformation process for three hierarchies, it is obvious that it can be extended to four, five or many hierarchies. However, the problem will be increasingly complex and non-linear as the number of hierarchy increases.

Example 3.3. Consider the following three-level problem solve by Anandalingam [1988]:

$$\max_{x1} f_1 = 7x_1 + 3x_2 + 4x_3 \qquad \text{(1}^{\text{st}}\text{ level)}$$

where x_2, x_3 solve:

$$\max_{x2} f_2 = x_2 \qquad \text{(2}^{\text{nd}}\text{ level)}$$

where x_3 solves:

$$\max_{x3} f_3 = x_3 \qquad \text{(3}^{\text{rd}}\text{ level)}$$

subject to:

$$\begin{aligned}
&x_1 + x_2 + x_3 \le 3 \\
&x_1 + x_2 - x_3 \ge 1 \\
&x_1 + x_2 + x_3 \ge 1 \\
&-x_1 + x_2 + x_3 \le 1 \\
&x_3 \le 0.5 \\
&x_1, x_2, x_3 \ge 0
\end{aligned}$$

Applying the KKT conditions to the lower two levels, we have:

$$\max_{x1} f_1 = 7x_1 + 3x_2 + 4x_3$$

where x_2, x_3 solve:

$$\max_{x2} f_2 = x_2$$

subject to:

$$\begin{aligned}
&\boldsymbol{x} \in \boldsymbol{Y}\\
&w_1 + w_2 - w_3 + w_4 + w_5 = 1\\
&w_1\,(x_1 + x_2 + x_3 - 3) = 0\\
&w_2\,(-x_1 - x_2 + x_3 + 1) = 0\\
&w_3\,(-x_1 - x_2 - x_3 + 1) = 0\\
&w_4\,(-x_1 + x_2 + x_3 - 1) = 0\\
&w_5\,(x_3 - 0.5) = 0\\
&w_1, w_2, w_3, w_4, \mathrm{w}_5 \geq 0
\end{aligned}$$

Applying the KKT conditions again to the above problem, we have:

$$\max f_1 = 7x_1 + 3x_2 + 4x_3$$

subject to:

$$\begin{aligned}
&\boldsymbol{x} \in \boldsymbol{Y}\\
&w_1 + w_2 - w_3 + w_4 + w_5 = 1\\
&w_1\,(x_1 + x_2 + x_3 - 3) = 0\\
&w_2\,(-x_1 - x_2 + x_3 + 1) = 0\\
&w_3\,(-x_1 - x_2 - x_3 + 1) = 0\\
&w_4\,(-x_1 + x_2 + x_3 - 1) = 0\\
&w_5\,(x_3 - 0.5) = 0\\
&w_1, w_2, w_3, w_4, w_5 \geq 0\\
&v_1\,(x_1 + x_2 + \mathrm{x}_3 - 3) = 0\\
&v_2\,(-x_1 - x_2 + x_3 + 1) = 0\\
&v_3\,(-x_1 - x_2 - x_3 + 1) = 0\\
&v_4\,(-x_1 + x_2 + x_3 - 1) = 0\\
&v_5\,(x_3 - 0.5) = 0\\
&(v_1 - v_2 - v_3 + v_4) + w_1\,\lambda_1 - w_2\,\lambda_2 - w_3\,\lambda_3 + w_4\,\lambda_4 = 1\\
&v_1, v_2, v_3, v_4, \lambda_1, \lambda_2, \lambda_3, \lambda_4 \geq 0
\end{aligned}$$

Let:

$$\begin{aligned}
&u_1 = v_1 + w_1\,\lambda_1\\
&u_2 = v_2 + w_2\,\lambda_2\\
&u_3 = v_3 + w_3\,\lambda_3\\
&u_4 = v_4 + w_4\,\lambda_4
\end{aligned}$$

then the one-level problem becomes:

$$\max f_1 = 7x_1 + 3x_2 + 4x_3$$

subject to:

$$\begin{aligned}
&\boldsymbol{x} \in \mathbf{Y}\\
&w_1 + w_2 - w_3 + w_4 + w_5 = 1
\end{aligned}$$

$$w_1 (x_1 + x_2 + x_3 - 3) = 0$$
$$w_2 (-x_1 - x_2 + x_3 + 1) = 0$$
$$w_3 (-x_1 - x_2 - x_3 + 1) = 0$$
$$w_4 (-x_1 + x_2 + x_3 - 1) = 0$$
$$w_5 (x_3 - 0.5) = 0$$
$$w_1, w_2, w_3, w_4, w_5 \geq 0$$
$$v_1 (x_1 + x_2 + x_3 - 3) = 0$$
$$v_2 (-x_1 - x_2 + x_3 + 1) = 0$$
$$v_3 (-x_1 - x_2 - x_3 + 1) = 0$$
$$v_4 (-x_1 + x_2 + x_3 - 1) = 0$$
$$v_5 (x_3 - 0.5) = 0$$
$$u_1 - u_2 - u_3 + u_4 = 1$$
$$u_1 = v_1 + w_1 \lambda_1$$
$$u_2 = v_2 + \underline{w_2} \lambda_2$$
$$u_3 = v_3 + w_3 \lambda_3$$
$$u_4 = v_4 + w_4 \lambda_4$$
$$v_1, v_2, v_3, v_4, \lambda_1, \lambda_2, \lambda_3, \lambda_4, u_1, u_2, u_3, u_4 \geq 0$$

The above problem can be transformed into a mixed-integer problem by using the Fortuny-Amat and McCarl transformation [1981], which was discussed in Section 3.1.1. Using the LINGO programming code [1992] and with $M = 10{,}000$, this mixed-integer programming was solved. The solution obtained is $f_1^* = 8.5$ at $(x_1^*, x_2^*, x_3^*) = (1.5, 0, 0.5)$, with $f_2^* = 0$, $f_3^* = 0.5$. We also solved the problem with M = 100,000, 1,000,000, 10,000,000, and 100,000,000 and the solution remains unchanged.

3.2.3 Simplex-cutting-plane Algorithm

Bard [1984b] proposed a cutting-plane algorithm employing a vertex-search procedure to solve the linear TLPP. To obtain the algorithm, Bard [1984b] first considered the following two-level problem obtained using the KKT conditions:

$$\max_{x1} f_1(\boldsymbol{x}) = \boldsymbol{c}_{11}{}^T \boldsymbol{x}_1 + \boldsymbol{c}_{12}{}^T \boldsymbol{x}_2 + \boldsymbol{c}_{13}{}^T \boldsymbol{x}_3 \tag{3.19}$$

where x_2, x_3 solve:

$$\max_{x2,\, x3} f_2(\boldsymbol{x}_1, \boldsymbol{x}_2, \boldsymbol{x}_3) = \boldsymbol{c}_{22}{}^T \boldsymbol{x}_1 + \boldsymbol{c}_{22}{}^T \boldsymbol{x}_2 + \boldsymbol{c}_{23}{}^T \boldsymbol{x}_3$$

subject to:

$$\boldsymbol{A}_2 \boldsymbol{x}_2 + \boldsymbol{A}_3 \boldsymbol{x}_3 \leq \boldsymbol{b} - \boldsymbol{A}_1 \boldsymbol{x}_1$$
$$\boldsymbol{u}^T \boldsymbol{A}_3 = \boldsymbol{c}_{33}$$
$$\boldsymbol{u}^T (\boldsymbol{A}_2 \boldsymbol{x}_2 + \boldsymbol{A}_3 \boldsymbol{x}_3 + \boldsymbol{A}_1 \boldsymbol{x}_1{}^0 - \boldsymbol{b}) = 0$$
$$\boldsymbol{x}_2, \boldsymbol{x}_3, \boldsymbol{u} \geq 0$$

Attaching an appropriately large weight K to the complementary equation and subtract this equation from the second-level objective function, we have:

$$\max_{x1} f_1(x) = c_{11}{}^T x_1 + c_{12}{}^T x_2 + c_{13}{}^T x_3 \tag{3.20}$$

where x_2, x_3 solve:

$$\max_{x2,x3,u} f_2(x_1, x_2, x_3) = c_{22}{}^T x_1 + c_{22}{}^T x_2 + c_{23}{}^T x_3 - K\, u^T (A_2 x_2 + A_3 x_3 + A_1 x_1{}^0 - b)$$

subject to:

$$A_2 x_2 + A_3 x_3 \le b - A_1 x_1$$
$$u^T A_3 = c_{33}$$
$$x_2, x_3, u \ge 0$$

Bard proved that a necessary condition that x^* solves the linear TLPP is that there exists a u^*, w^*, and $v^* \in R^m$, such that (x^*, u^*, w^*, v^*) is feasible for the problem:

$$\max f_1(x_1, x_2, x_3) = c_{11}{}^T x_1 + c_{12}{}^T x_2 + c_{13}{}^T x_3 \tag{3.21}$$

subject to:

$$A_1 x_1 + A_2 x_2 + A_3 x_3 \le b$$
$$u^T A_3 = c_{33}$$
$$w^T A_2 = c_{22}$$
$$w^T A_3 = c_{23}$$
$$K u - v + w = 0$$
$$v^T (A_1 x_1 + A_2 x_2 + A_3 x_3 - b) = 0$$
$$u^T (A_1 x_1 + A_2 x_2 + A_3 x_3 - b) = 0$$
$$x_1, x_2, x_3, u, w, v \ge 0$$

where u , v and w are KKT multipliers associated with the constraint sets, and K is a sufficiently large finite constant.

Based on Equation 3.21 and an upper bound for the objective function f_1, Bard developed the following simplex-cutting plane algorithm with simplex vertex enumeration:

Step 1. Solve Equation 3.21 to obtain the trial point $\underline{\underline{x}} = (\underline{\underline{x}}_1, \underline{\underline{x}}_2, \underline{\underline{x}}_3)$, and a tight upper bound on f_1. Go to Step 2.

Step 2. Fix x_1 at $\underline{\underline{x}}_1$ and solve the second-level problem only in Equation 3.7 to obtain $\underline{x}_{[1]}$. If $\underline{\underline{x}} = \underline{x}_{[1]}$ or if $f_1(\underline{x}_{[1]}) = f_1(\underline{\underline{x}})$, stop. Otherwise, let $\underline{f}_1 = f_1(\underline{x}_{[1]})$ as a lower bound and go to Step 3.

Step 3. Search for a vertex adjacent to $\underline{x}_{[k]}$ that both lies in the first-level inducible region and is in the direction of increase of f_1. If none is found, go to Step 4. Otherwise, move to this vertex and repeat the search until a local optimum is reached, call it x^* and update the lower bound by setting $\underline{f}_1 = f_1(x^*)$.

Step 4. Add the cut $f_1 \leq \underline{f}_1 + \varepsilon$ to the feasible set of Equation 3.21 and solve to obtain $\underline{\underline{x}}$. $\varepsilon > 0$ is a sufficiently small constant so that no unexplored vertices of the original constraint set are eliminated.

Step 5. Test to see whether $\underline{\underline{x}}$ is in the region by fixing x_1 at $\underline{\underline{x}}_1$, and solve the second-level problem only in Equation 3.7. If the test is positive, put $\underline{x}_{[k+1]} = \underline{\underline{x}}$, eliminate the cut and go to Step 3. Otherwise, search adjacent vertices in the direction of increasing $\underline{x}_{[k+1]}$ in the first-level inducible region. Repeat search if necessary starting from each of the other multiple solutions obtained at Step 4 until such a point is found; then go to Step 3. If the search fails, stop and x^* is the global optimum.

The following problem solved by Bard [1984b] is used to illustrate the approach.

Example 3.4.

$$\max_{x1} f_1 = -4x_1 + 2x_2 - 5x_3 \qquad (1^{st} \text{ level})$$

where x_2, x_3 solve:

$$\max_{x2} f_2 = -x_2 + 4x_3 \qquad (2^{nd} \text{ level})$$

where x_3 solves:

$$\max_{x3} f_3 = 2x_3 \qquad (3^{rd} \text{ level})$$

subject to:

$$\begin{aligned} &2x_2 - x_3 \geq 2 \\ &-3x_1 + x_2 - x_3 \geq -12 \\ &-3x_2 - x_3 \geq -24 \\ &x_1 \geq 2 \\ &-x_3 \geq -6 \\ &x_1, x_2, x_3 \geq 0 \end{aligned}$$

The corresponding Equation 3.21 is obtained as:

$$\max f_1 = 7x_1 + 3x_2 + 4x_3$$

subject to:

$$\begin{aligned} &2x_2 - x_3 \geq 2 \\ &-3x_1 + x_2 - x_3 \geq -12 \\ &-3x_2 - x_3 \geq -24 \end{aligned}$$

$$
\begin{aligned}
& x_1 \geq 2 \\
& -x_3 \geq -6 \\
& u_1 + u_2 + u_3 + u_5 = 2 \\
& w_1 + w_2 + w_3 + w_5 = 4 \\
& 2w_1 + w_2 - 3w_3 = 1 \\
& K u_1 - v_1 + w_1 = 0 \\
& K u_2 - v_2 + w_2 = 0 \\
& K u_3 - v_3 + w_3 = 0 \\
& K u_4 - v_4 + w_4 = 0 \\
& K u_5 - v_5 + w_5 = 0 \\
& u_1 (2x_2 - x_3 - 2) = 0 \\
& u_2 (-3x_1 + x_2 - x_3 + 12) = 0 \\
& u_3 (-3x_2 - x_3 + 24) = 0 \\
& u_4 (x_1 - 2) = 0 \\
& u_5 (-x_3 + 6) = 0 \\
& v_1 (2x_2 - x_3 - 2) = 0 \\
& v_2 (-3x_1 + x_2 - x_3 + 12) = 0 \\
& v_3 (-3x_2 - x_3 + 24) = 0 \\
& v_4 (x_1 - 2) = 0 \\
& v_5 (-x_3 + 6) = 0 \\
& \boldsymbol{u}, \boldsymbol{v}, \boldsymbol{w}, \boldsymbol{x} \geq \boldsymbol{0}
\end{aligned}
$$

and K is a sufficiently large finite constant.

Table 3.1. Decisions and objective functions, Example 3.4 [Bard 1984]

Vertex	(x_1, x_2, x_3)	F_1	f_2	F_3
Global optimum, A	(14/3, 1, 0)	-50/3	-1	0
Local optimum, B	(2, 4, 6)	-30	20	12
Feasible point, C	(10/3, 4, 6)	-106/3	20	12
High point, D	(2, 8, 0)	8	-8	0
(3.15) solution, E	(2, 1, 0)	-6	-1	0

The following discussions follow the solution carried out by Bard [1984b].

Let $K = 100$ and $\varepsilon = 1$ and apply Step 1 to the above problem, the solution obtained is $\underline{\underline{x}} = (\underline{\underline{x}}_1, \underline{\underline{x}}_2, \underline{\underline{x}}_3) = (2, 1, 0)$, which is Point E in Table 3.1 and this solution is not in level 1 inducible region.

Fixing x_1 at 2 and solve the level 2 problem in Equation 3.7, we obtain point B or $\underline{\underline{x}}_{[1]} = (2, 4, 6)$, which belongs to the level 1 inducible region. The search at Step 3 leads to the conclusion that point B is a local solution because there are no adjacent vertices. At Step 4, the cut $f_1 = -4x_1 + 2x_2 - 5x_3 \leq -30 + \varepsilon$ is added to the constraint region of the transformed formulation and obtains a new point $\underline{\underline{x}} = (4.57,$

6.43, 4.71). This point $\underline{\underline{x}}$ is not in the level 1 inducible region; therefore, at Step 5 a search of adjacent vertices must be conducted to find an element of the level 1 inducible region that intersects the cut, should one exist. The search leads to the point $\underline{\underline{x}}_{[2]}$ = (3.65, 3.05, 4.1) that lies along the line from point A, which is (14/3, 1, 0) to point C, which is (10/3, 4, 6), on the level 1 inducible region. Returning to Step 3 brings us to point A, which turns out to be the local optimum x^* = (14/3, 1, 0). A second cut must now be added but the algorithm terminates at Step 5 when no new point in the level 1 inducible region can be found that produces a value of f_1 larger than -50/3, the best lower bound.

3.2.4 Penalty-function Approach

Anandalingdam and Apprey [1991] proposed an algorithm that combines k^{th}-best algorithm with penalty function to solve the linear three-level problem. First, the following first-level problem is solved:

$$\max f_1(x) = c_{11}^T x_1 + c_{12}^T x_2 + c_{13}^T x_3 \tag{3.22}$$

subject to:

$$A_1 x_1 + A_2 x_2 + A_3 x_3 \leq b$$
$$x_1, x_2, x_3 \geq 0$$

Let $(x_1^{1*}, x_2^{1*}, x_3^{1*})$ represent this solution. For a given x_1^{1*}, the two-lower-level problem becomes:

$$\max_{x2} f_2(x) = c_{21}^T x_1^{1*} + c_{22}^T x_2 + c_{23}^T x_3 \tag{3.23}$$

where x_3 solves:

$$\max_{x3} f_3(x) = c_{31}^T x_1^{1*} + c_{32}^T x_2 + c_{33}^T x_3$$

subject to:

$$A_2 x_2 + A_3 x_3 \leq b - A_1 x_1^{1*}$$
$$x_2, x_3 \geq 0$$

The global optimal solution for Equation 3.23 is obtained through the algorithm listed in Section 3.1.2. Suppose the solution is (x_2^1, x_3^1), and if $x_2^1 = x_2^{1*}$ and $x_3^1 = x_3^{1*}$, then $(x_1^{1*}, x_2^{1*}, x_3^{1*})$ is the optimal solution for the overall problem. Otherwise, the second best solution $(x_1^{2*}, x_2^{2*}, x_3^{2*})$ is obtained from Equation 3.22, the process is repeated. The suggested algorithm can be summarized as follows:

Step 1. Solve Equation 3.22 and let the solution be represented by $x^{1*} = (x_1^{1*}, x_2^{1*}, x_3^{1*})$. Go to Step 2.

Step 2. Solve Equation 3.23 by the algorithm in Section 3.1.2 with $\boldsymbol{x}_1$ given as $\boldsymbol{x}_1{}^{1*}$ and let this solution be represented by $\boldsymbol{x}^1 = (\boldsymbol{x}_1{}^{1*}, \boldsymbol{x}_2{}^1, \boldsymbol{x}_3{}^1)$. Go to Step 3.

Step 3. If $\boldsymbol{x}^1 = \boldsymbol{x}^{1*}$, the optimality is reached and stop. Otherwise, go to Step 4.

Step 4. Let $\boldsymbol{x}^{2*} \in X_v$ be the adjacent point to $\boldsymbol{x}^{1*}$. Set $\boldsymbol{x}^{1*} = \boldsymbol{x}^{2*}$, go to Step 2.

The proposed procedure essentially follows the kth-best algorithm to search for k^{th} extreme point, and the optimal solution is reached if the adjustments between the first-level (leader) and the two lower levels (followers) are completed.

3.3 Non-linear Multi-level Programming

At least theoretically, most of the transformation approaches discussed early can be extended or modified for solving non-linear MLPPs. In fact, some of them were developed for non-linear problems. One example is the branch-and-bound approach, which was developed for linear or linear-quadratic BLPPs [Bard and Moore 1990]. Bard and Moore further pointed out that their approach could be used to solve non-linear problems if a more general computer code such as GRG2 was used. However, general non-linear MLPPs are much more difficult to solve. Most of the examples in the literature are concerned with quadratic problems. The interactive fuzzy approach to be discussed in later chapters appears to have more potential for solving general non-linear MLPPs. An added advantage of the interactive approach is the realistic representation of a basically vague and not well defined complex hierarchy system.

To illustrate the approaches, several non-linear algorithms will be discussed in this section. Most of the examples are quadratic problems, which are easier to solve than the more general non-linear problems.

3.3.1 Sequential Linear-quadratic Complementary Algorithm with Branch-and-bound

The hybrid algorithm, which was used to solve BLPP in Section 2.3.2, and which was formed by the combined use of a sequential solution method to solve the LCP and a branch-and-bound search, was again used for solving linear-quadratic BLPPs after modifications [Judiice and Faustino 1994]. Consider the linear-quadratic problem:

$$\min_{x1} f_1(x_1, x_2) = c_{11}{}^T x_1 + c_{12}{}^T x_2 \tag{3.24}$$

subject to:

$$E_1 x_1 + E_2 x_2 \geq e$$
$$x_1 \geq 0$$

where x_2 solves:

$$\min_{x2} f_2(x_1, x_2) = c_{22}{}^T x_2 + x_1{}^T S\, x_2 + ½\, x_2\, Q\, x_2 \tag{3.25}$$

subject to:

$$A_1 x_1 + A_2 x_2 \geq b$$
$$x_2 \geq 0$$

where E_1 and E_2 are $\ell \times n_1$ and $\ell \times n_2$ dimensional matrices, respectively; A_1 and A_2 are $m \times n_1$ and $m \times n_2$ dimensional matrices, respectively; S and Q are $n_1 \times n_2$ and $n_2 \times n_2$ dimensional matrices, respectively; x_1 and x_2 are n_1 and n_2 dimensional vectors, respectively; and e and b are ℓ- and m dimensional vectors, respectively. In addition to the second-level constraint set, the above problem also has first level constraint set. When the first-level constraint set is absent, which is the case for the problems discussed before, several authors have shown that the rational set is a connected union of faces of the second-level polytope (see Figure 2.1). This conclusion does not hold for the above problem represented by Equations 3.24 and 3.25. The polyhedron defined by the first and second-level constraints is called the relaxed feasible region. This region is generally assumed nonempty and bounded. If, for each fixed x_1, the constraint set of the follower's problem in x_2 is bounded, then we can apply the KKT conditions to Equations 3.24 and 3.25 and obtain the following minimum linear complementary problem [Judice and Fanstino 1994]:

$$\min c_{11}{}^T x_1 + c_{12}{}^T x_2 \tag{3.26}$$

subject to:

$$\alpha = A_1 x_1 + A_2 x_2 - b$$
$$\beta = c_{22} + S^T x_1 + Q\, x_2 - A_2{}^T u$$
$$\Upsilon = -e + E_1 x_1 + E_2 x_2$$
$$x_2{}^T \beta = u^T \alpha = 0$$
$$x_1, x_2, u, \alpha, \beta, \Upsilon \geq 0$$

Introducing a parameter λ and replacing the objective function of the above equation by the inequality:

$$c_{11}{}^T x_1 \ + \ c_{12}{}^T x_2 \leq \lambda$$

After introducing the slack variable v_0 into the above inequality, Equation 3.26, in matrix form, can be represented by the following linear complementary problem (LCP):

$$\begin{bmatrix} \alpha \\ \beta \\ \gamma \\ v_0 \end{bmatrix} = \begin{bmatrix} -b \\ c_{22} \\ -e \\ \lambda \end{bmatrix} + \begin{bmatrix} 0 & A_2 & A_1 \\ -A_2{}^T & Q & S^T \\ 0 & E_2 & E_1 \\ 0 & -c^T{}_{12} & -c^T{}_{11} \end{bmatrix} \begin{bmatrix} u \\ x_2 \\ x_1 \end{bmatrix} \tag{3.27}$$

$x_2{}^T\beta = u^T\alpha = 0$
$x_1, x_2, u, \alpha, \beta, \gamma \geq 0$

The algorithm, which is essentially the same as that for linear BLPP discussed in Section 2.3.2, consists of solving a sequence of LCPs with decreasing values of the real number λ. The matrix equation, Equation 3.27, can be put into the same form as Equation 2.12 in Chapter 2:

$$w = q + Mz + Nx \tag{3.28}$$
$$w \geq 0, \ z \geq 0, \ \underline{x} \geq 0$$
$$z_i \, w_i = 0, \ i = 1, 2, 3, \ldots, n_2 + m$$

where

$$z = \begin{bmatrix} u \\ x_2 \end{bmatrix}, \qquad w = \begin{bmatrix} \alpha \\ \beta \\ \gamma \\ v_0 \end{bmatrix} \qquad q = \begin{bmatrix} -b \\ c_{22} \\ -e \\ \lambda_k \end{bmatrix}$$

$$M = \begin{bmatrix} o & A_2 \\ -A_2{}^T & Q \\ o & E_2 \\ 0 & -c_{12}{}^T \end{bmatrix}, \qquad N = \begin{bmatrix} A_1 \\ S^T \\ E_1 \\ -c_{11}{}^T \end{bmatrix}$$

where the subscript k in λ_k, $k = 0,1,2,\ldots$ is a sequence of decreasing values of the parameter k and also represents the number of iterations in the actual computational scheme.

The linear-quadratic BLPP is solved by first solving the above LCP problem using a branch-and-bound algorithm with the help of a modified reduced gradient technique due to Al-Khayyal [1987]. Algorithms SIMPMRG and MAXVAR were used. The investigators solved problems with a maximum of 70 and 50 decision variables for the upper and lower DMs, respectively. The problems solved have up to 12 and 36 constraints for the upper and lower levels, respectively. The test problems were generated according to the recommendations of Hansen *et al.* [1990] and Edmunds and Bard [1991]. The densities of the matrices $A = \{A_1, A_2]$ and $E = [E_1, E_2]$ are 8% and the densities for Q are 5%. All the test problems were also solved by the branch-and-bound approach due to Bard and Moore[1990]. These authors discovered that, for small and medium scale linear-quadratic problems, the proposed approach is again consistently more efficient compared to the branch-and-bound approach proposed by Bard and Moore [1990]. The efficiency of the current approach becomes more pronounced as the dimension of the problem increases. Furthermore, the SLCP approach is robust and is quite efficient at finding global optimum.

3.3.2 Steepest-descent Approach

Savard and Gauvin [1994] developed a steepest-descent approach to solve the non-linear BLPP. A first order necessary optimality condition for non-linear BLPP was first obtained based on certain assumptions such as well posed *etc.* It was shown that the directional derivative could be obtained by solving a quadratic BLPP. In the following, we shall follow the presentation of Savard and Gauvin [1994].

Consider the non-linear BLPP:

$$\min_x F(x, y) \qquad \text{(upper level)} \quad (3.29)$$

where y is obtained by solving the lower-level problem:

$$\min_y f_0(x, y) \qquad \text{(lower level)} \quad (3.30)$$

subject to:

$$f_i(x, y) \le 0$$
$$f_j(x, y) = 0$$

where $i = 1, 2,\ldots, p$ and $j = 1, 2,\ldots, q$. In other words, there are p inequality and q equality constraints. x and y are n and m-dimensional vectors, respectively. Since y is obtained based on the already decided x, we can write $y(x)$. The above non-linear BLPP is assumed to obey three assumptions. Two of the assumptions are listed in the following and will be used in the algorithms for numerical solutions:

1. for any x, the optimal solution $y(x)$ for the lower problem is unique;
2. the vectors

   $$\nabla_y f_s(x, y(x))$$

   are linearly independent with $s \in r \cup q$, where r is the number of effective inequality constraints. In other words, r is the number of inequality constraints where the current solution point $(x, y(x))$ is on these or against these inequality constraints.

Based on these assumptions, a first order necessary condition was obtained and the steepest-descent direction for the non-linear BLPP is given by any optimal solution of the following linear-quadratic BLPP:

$$\min_z \nabla_x F(x, y(x))\, z + \nabla_y F(x, y(x))\, w(x, z) \qquad \text{(upper level)} \quad (3.31)$$

where $w(x, z)$ is an m-dimensional vector and is obtained by solving the lower-level quadratic problem:

$$\min_w (z^T, w^T)\, \nabla^2_{(x,y)} L(x, y(x), \lambda(x))\, (z, w) \qquad \text{(lower level)} \quad (3.32)$$

subject to:

$$\begin{aligned}
&\nabla_y f_i\, (x, y(x))w \le -\nabla_x f_i\, (x, y(x))z, \;\; i\in r\\
&\nabla_y f_j\, (x, y(x))w = -\nabla_x f_j\, (x, y(x))z, \;\; j\in q\\
&\nabla_y f_0\, (x, y(x))w = -\nabla_x f_0\, (x, y(x))z + \nabla_x L(x, y(x), \lambda(x))z\\
&-1 \le z_k \le 1,\; k=1, 2,\ldots, n
\end{aligned}$$

where $\nabla^2_{(x,y)} L(x, y(x), \lambda(x))$ in the objective function of Equation 3.32 is the Hessian matrix of the following Lagrangian of the lower objective function of the non-linear BLPP

$$L(x, y(x), \lambda(x)) = f_0\, (x, y) + \textstyle\sum_{s\in r\cup q} \lambda_s f_s\, (x,y) \qquad (3.33)$$

z and w are n and m dimensional vectors, respectively. The expression $\nabla_y f_i(x, y(x))$ is the partial derivative of f_i with respect to y. Notice that the upper-level objective function is linear while the lower-level objective function is quadratic. The above linear-quadratic BLPP, which is represented by Equations 3.31 and 3.32, can be solved by using either the branch-and-bound approach of Bard and Moore [1990] or the hybrid SLCP algorithm due to Judice and Faustino [1992]. Both of these approaches were discussed in the previous sections.

Example 3.5. This example follows the work of Savard and Gauvin [1994]. Consider the non-linear BLPP:

$$\min_x (x-1)^2 + 2y_1^2 - 2x$$

where y solves:

$$\min_y (2y_1 - 4)^2 + (2y_2 - 1)^2 + xy_1$$

subject to:

$$\begin{aligned}
4x + 5y_1 + 4y_2 &\le 12\\
-4x - 5y_1 + 4y_2 &\le -4\\
4x - 4y_1 + 5y_2 &\le 4\\
-4x + 4y_1 + 5y_2 &\le 4
\end{aligned}$$

$$x \geq 0, y_1 \geq 0, y_2 \geq 0$$

Savard and Gauvin [1994] obtained the initial rational point by first solving the above problem with the lower-level objective function ignored. Or, solving the following one-level problem:

$$\min_x (x-1)^2 + 2y_1^2 - 2x$$

subject to:

$$\begin{aligned} 4x + 5y_1 + 4y_2 &\leq 12 \\ -4x - 5y_1 + 4y_2 &\leq -4 \\ 4x - 4y_1 + 5y_2 &\leq 4 \\ -4x + 4y_1 + 5y_2 &\leq 4 \\ x \geq 0, y_1 \geq 0, y_2 &\geq 0 \end{aligned}$$

The optimal solution for this one-level problem is $(x, y_1, y_2)=(4/3, 1/3, 0)$. Now the lower-level problem, with x given as 4/3, is solved and the solution is: $(x, y_1, y_2)=(4/3, 158/123, 5/82)$, which is the rational point $(x, y(x))$. Only the first inequality constraint which lies on this rational point, or this rational point only against this first constraint, and thus $r = 1$. The corresponding Lagrange multiplier for this first constraint is $\lambda_1 = 72/82$. The linear-quadratic BLPP based on this rational point can be obtained by differentiation of the original non-linear BLPP and the Lagrangian equation and by using Equations 3.31-3.33:

$$\min_z -4/3\, z + 632/123\, w_1$$

$$\min_w 2\, z\, w_1 + 8\, w_1^2 + 8\, w_2^2$$

subject to:

$$\begin{aligned} 5\, w_1 + 4\, w_2 &\leq -4\, z \\ -5w_1 - 4\, w_2 &= 4z \\ -1 \leq z &\leq 1 \end{aligned}$$

The solution for this problem is $(z, w_1, w_2) = (1, -22/41, -27/82)$. This solution is also the steepest-descent direction. If this direction is non-negative, then a local minimum is obtained. Stop the iteration. If, on the other hand, this direction is negative, the problem can be further improved along this steepest-descent direction and a step size should be chosen. The solution (1, -22/41, -27/82) is in the negative direction. The size of the step to move in this direction was obtained by finding the minimum along this direction as follows:

$$\min_{t>0} F((4/3, 158/123, 5/82) + t\,(1, -22/41, -27/82))$$

where t is the step size. The value of t should be larger than 0 but should be small enough so that the constraints of the problem are not violated. The value of t which gives the minimum of the above expression, at the same time, does not violate the constraints is t^*=5/27. Substituting t into the following expression, we have:

$$(4/3, 158/123, 5/82) + t\,(1, -22/41, -27/82) = (41/27, 32/27, 0)$$

At this newly improved point (x, y_1, y_2)=(41/27, 32/27, 0), the first and the last constraints are active, or r = 2. The Lagrange multipliers are λ_1 =4 and λ_7 = 0. Again, obtain the linear-quadratic BLPP at this point:

$$\min_z -26/27\, z + 128/27\, w_1$$

$$\min_w 2\, z\, w_1 + 8\, w_1^2 + 8\, w_2^2$$

subject to:

$$\begin{aligned}
&5\, w_1 + 4\, w_2 \le -4\\
&w_2 \ge 0\\
&-5\, w_1 - 4\, w_2 = 4\, z\\
&-1 \le z \le 1
\end{aligned}$$

The solution of this problem is (z, w_1, w_2) = (1, -4/5, 0) with negative direction. Use the same approach, the step size obtained: t = 10/27 and the newly improved point: (x, y_1, y_2)= (17/9, 8/9, 0). The first, third and seven constraints lie on this point. Thus, r = 3 and λ_1 = 1.4, λ_3 = 0, and λ_7 = 1.6. Form the linear-quadratic BLPP on this point, we have:

$$\min_z -2/9\, z + 32/9\, w_1$$

$$\min_w 2\, z\, w_1 + 8\, w_1^2 + 8\, w_2^2$$

subject to:

$$\begin{aligned}
&5\, w_1 + 4\, w_2 \le -4\, z\\
&-4\, w_1 + 5\, w_2 \le -4\, z\\
&w_2 \ge 0\\
&-7\, w_1 - 4\, w_2 = -8/9\, z\\
&-1 \le z \le 1
\end{aligned}$$

The optimal solution is (z, w_1, w_2) = (0, 0, 0) with the direction non-negative or the direction is zero. Thus, the newly obtained point (x, y_1, y_2)=((17/9, 8/9, 0) is a local optimum. In fact, this solution is the global optimum of the non-linear BLPP.

3.3.3 Evolutionary Approach: Genetic Algorithms

Although reasonably effective branch-and-bound approaches such as that due to Bard and Moore [1992], Judice and Faustino [1992], and Hansen *et al.* [1992] have been developed, large problems still have computational limitations. Evolutionary techniques, which are essentially heuristic local search approaches, are ideally suited for solving NP-hard problems and can extend the range of solvable BLPP and MLPP problems considerably. However, very little search has been carried out in this direction.

Gendreau *et al.* [1996] used a hybrid tabu search and ascent algorithm to approximately solve the linear BLPP by utilizing both the primal and the dual versions of the problem, with the help of the algorithm of Hansen *et al.* [1992]. These investigators, following the approach of Anandalingam and White [1990], reduced the bi-level problem to one-level by incorporating the complementary slackness into the leader's objective function after multiply by a large positive number, which forms the penalty. The algorithm performed well for solving a large set of difficult test problems involving up to 200 variables and 200 constraints. Optimum was obtained for a majority of the problem and, on the average, the relative errors are extremely small. Wen and Huang [1996] investigated the use of tabu search to solve the mixed-integer BLPP.

Liu [1998] used genetic algorithms to obtain the Stackelberg-Nash equilibrium for the non-linear BLPP and non-linear multi-level problem with multiple followers. To avoid the problems when the follower's rational reaction set is not a singleton, Liu assumed that the problems considered have no multiple values in the rational reaction set of the follower's problem.

In Chapter 1, the arrays $(y_1^*, y_2^*, \ldots, y_p^*)$ and $(x^*, y_1^*, y_2^*, \ldots, y_p^*)$ were defined as the Nash and Stackelberg-Nash equilibrium. The non-linear BLDPP, which was also discussed in Chapter 1, is repeated in the following:

$$\max_x F(x, y_1, y_2, \ldots, y_p) \tag{3.34}$$

subject to:

$$G(x) \leq 0$$

where each y_i, I = 1, 2, …, p, solves:

$$\max_{xi} f_i(x, y_1, y_2, \ldots, y_p) \tag{3.35}$$

subject to:

$$g(x, y_1, y_2, \ldots, y_p) \leq 0$$

where x is the control of the leader and $y_1, y_2, \ldots, y_p$ are the controls of the p followers, and F and f_i are the objective functions of the leader and the ith follower, respectively.

To solve for the Stackelberg-Nash equilibrium or to solve the above non-linear BLDPP problem, define the other followers as:

$$y_{-i} = (y_1, y_2, \ldots, y_{i-1}, y_{i+1}, \ldots, y_p),\ \mathrm{i} = 1, 2, \ldots, p \tag{3.36}$$

After the decision revealed by the leader and the other followers y_{-i}, the optimal reaction of the ith follower must be obtained based on these already decided decisions. In other words, the Stackelberg-Nash equilibrium for the ith follower is imbedded in these already decided decisions, or $y_i = h(x, y_{-i})$. This implicit function cannot be solved easily, especially for non-linear problems. Genetic algorithms are the ideal approach for obtaining solution of this implicit problem.

Genetic algorithms [Holland 1975, Goldberg 1989, Davis 1991] were developed by Holland in the 1960s. The algorithms have been applied successfully in various different fields, principally in machine learning and optimization. We shall not go into the details of the algorithm. The reader who is not familiar with the technique can consult the references listed above.

The algorithm is based on the mechanisms of natural selection, evolution, and genetics. The operation is based, in general, on the manipulation of strings of binary digits called chromosomes. The principal operations, in its simplest form, consist of reproduction, crossover, and mutation. The reproduction process can be conducted by the use of a biased roulette wheel, whose slots have different sizes proportional to the fitness values of the members.

Since the values for each DM in the Stackelberg-Nash equilibrium is implicitly expressed as functions of other DMs, these values cannot be obtained easily. Liu [1998] used the genetic algorithms to search for these values. The steps of this search by the use of the genetic algorithms can be divided into: (1) initialization, (2) evaluation or the assignment of probability, (3) selection or reproduction, (4) crossover, and (5) mutation. For a detailed discussion, the reader is referred to the original literature [Liu 1998].

Liu solved various non-linear problems. To illustrate the usefulness of this algorithm, some of the solved problems and the obtained results are summarized in the following. The first problem solved is the following non-linear max-min problem:

$$\max F(x, y) = ((x_1 + y_1)(x_2 + y_2))/(1 + x_1y_1 + x_2y_2\}$$

subject to:

$$x_1^2 + x_2^2 \le 100$$
$$x_1 \ge 0$$
$$\mathrm{x}_2 \ge 0$$

$$\max f(x, y) = -F(x, y)$$

subject to:

$$0 \le y_1 \le x_1$$
$$0 \le y_2 \le x_2$$

The solution or the Stackelberg-Nash equilibrium obtained is $x^* = (7.0854, 7.0291)$ and $y^* = (7.0854, 0.0000)$ with $F(x^*, y^*) = 1.9760$. The solution was obtained after the program run 300 generations. The second problem solved by Liu is:

$$\max F(x, y_1, y_2) = (200 - y_{11} - y_{21})(y_{11} + y_{21}) + (160 - y_{12} - y_{22})(y_{12} + y_{22})$$

subject to:

$$\begin{aligned} &x_1 + x_2 + x_3 + x_4 \le 40 \\ &0 \le x_1 \le 10 \\ &0 \le x_2 \le 5 \\ &0 \le x_3 \le 15 \\ &0 \le x_4 \le 20 \end{aligned}$$

$$\min f_1(y_1) = (y_{11} - 4)^2 + (y_{12} - 13)^2$$

subject to:

$$\begin{aligned} &0.4y_{11} + 0.7y_{12} \le x_1 \\ &0.6y_{11} + 0.3y_{12} \le x_2 \\ &0 \le y_{11} \\ &y_{12} \le 20 \end{aligned}$$

$$\min f_2(y_2) = (Y_{21} - 35)^2 + (Y_{22} - 2)^2$$

subject to:

$$\begin{aligned} &0.4y_{21} + 0.7y_{22} \le x_3 \\ &0.6y_{21} + 0.3y_{22} \le x_4 \\ &0 \le y_{21} \\ &y_{22} \le 40 \end{aligned}$$

The following results were obtained after 600 generations by the genetic algorithms:

$$x^* = (7.05, 3.13, 11.93, 17.89),\ y_1^* = (0.26, 9.92),\ y_2^* = (29.82, 0.00)$$

with the objective function values:

$$F(x^*, y_1^*, y_2^*) = 6599.99,\ f_1(y_1) = 23.47,\ f_2(y_2) = 30.83$$

which compared favorably with the exact solution obtained by Bard [1988]:

$$x^* = (7.91, 4.37, 11.09, 16.63),\ y_1^* = (2.29, 10),\ y_2^* = (27.21, 0.00)$$

with the objective function: $F(x^*, y_1^*, y_2^*) = 6600$.

Liu also solved other examples. One of the examples contains sine functions in both the leader's and the three followers' objective functions. This is an especially difficult problem because of the multiple modalities of the object functions.

Although the numerical experiment is limited, it does show the usefulness of the evolutionary approaches in solving difficult non-linear MLPPs.

3.4 Discrete Bi-level Programming

There are two directions of research concerning the discrete problems. One is concerned directly with the numerical solution of discrete BLPP and the other is to investigate the interconnections between BLPPs and mixed-integer problems (MIP), and through which BLPP can be solved and *vice versa*. Moore and Bard [1990] developed a series of heuristics for solving the mixed-integer linear BLPP. They showed that several of the fathoming rules in solving MIP by the branch-and-bound approach cannot be directly applied for solving the mixed-integer linear BLPP. Bard and Moore [1992], by converting the leader's objective function into a parameterized constraint for the follower, developed a branch-and-bound approach for solving the zero-one integer BLPP. Wen and Yang [1990] developed a branch-and-bound algorithm for solving mixed 0-1 integer bi-level problem, where the upper level only has 0-1 integers and the variables of the bottom level is continuous. Jan and Chern [1994], using a parametric analysis approach, developed a non-linear integer BLPP algorithm, where the variables must be separable. Wen and Huang [1996] used tabu search to solve the mixed-integer linear BLPP.

In the second direction, Frangioni [1995] investigated the connections between BLPP and MIP. This connection leads to the use of BLPP algorithms as heuristic approaches for MIP. Savard *et al.* [1996] divided the discrete linear BLPP into three cases:

1. upper-level discrete and lower-level continuous (DC-BLPP);
2. upper-level continuous and lower-level discrete (CD-BLPP);
3. both upper and lower levels are integers (DD-BLPP).

These authors studied the interconnections between the linear BLPP and the discrete case and showed that DC-BLPP and DD-BLPP problems always have optimal solutions under hypothesis similar to the linear case. On the other hand, the existence of an optimal solution to the CD-BLPP seems to be a more difficult task. They also showed that DC-BLPP can always be reduced into a linear bi-level problem and DD-BLPP is equivalent to a linear three-level problem. However, this is not the case for the CD-BLPP.

Audet *et al.* [1997] studied the reformulation and imbedding of algorithms for mixed 0-1 linear programming problem and linear BLPP with or without some 0-1 variables at the first level. Their formulation of linear mixed 0-1 programming problem into a linear bi-level one does not require the introduction of a large number of finite constant. These investigators also showed that the classical branch-and-bound algorithm of Beale and Small for MIP [1965] is embedded in the algorithm of Hansen *et al.* [1992] for BLPP. Furthermore, these investigations of embedding appear to be very useful and the results can lead to generalization or specialization of tests from one algorithm to another.

3.5 Discussions

Because of the practical importance of the duo- or multi-ploy type decision-making process, many algorithms have been developed for solving this type of problems by using a variety of different approaches. In this and the last chapters, several of the representative approaches are introduced. Obviously, there are many other algorithms, it is impossible, nor suitable to introduce even most of the algorithms. Among the traditional optimization approaches, the several algorithms based on branch-and-bound appear to be most efficient. Most of these algorithms also borrow the idea of mixed integer to handle the complementary slackness term due to Fortuny-Amat and McCarl [1981]. The kth-best algorithm, although not computationally efficient, does give a reasonably clear picture of the problem involved. The main complication in solving the multi-ploy type problem is due to the fact that the rational reactions of the lower levels must be singletons in order to guarantee true optimum of the upper levels. It appears that the most promising future development so that more complicated practical problem can be solved is in the use of evolutionary approaches, which, although cannot guarantee true optimum, does avoid the problem of exponential explosion.

The algorithms introduced in this and the last chapters are not suited for solving interactive decision problems, where the mutual interactions between the different levels are more flexible and where the decision process is more in line with that of the hierarchy organizations such as the manufacturing or service industries. Furthermore, these algorithms are not very efficient even for the duo- or multi-ploy type systems. The approach by the use of fuzzy interactive decision making process, which is much more flexible to allow interactions between the different levels, will be discussed in the remainder of the book. Following the same approach, various examples will be given to illustrate the solution procedures.

Chapter 4

Possibility Theory and Knowledge Representation

Fuzzy set theory is a powerful tool for representing vague phenomena and for linguistic representation in modern computer. The basic concept has been applied in many different areas. For the convenience of later discussions, some of the fuzzy concepts are summarized in this chapter. However, we shall not discuss the most basic aspects and assume the reader has some familiarity with fuzzy set theory. For more details, the reader can consult the many excellent books on this subject. For example, the book by Klir and Folger [1988] and especially the papers by Zadeh, which has been collected in a single book [1987], can be consulted. The book by Lee and Zhu [1992, 1995] and the papers by Zadeh are more directly connected with this chapter. In fact, we follow these works closely during most of the discussions.

4.1 Possibility Theory

The concept of possibility proposed by Zadeh [1978, 1987] forms one of the most useful foundations in fuzzy set theory. By the use of this concept, concrete meanings can be established and used for the formation of fuzzy membership function. Furthermore, because fuzzy set is ideally suited for the representation of linguistic expressions, the frequently used linguistic expressions by management can be expressed in terms of membership functions. In order to use this concept in hierarchical decision-making, the meaning of possibility or possibility measure is first summarized in this section.

Possibility measure can be defined either based on confidence measure or based on fuzzy set [Lee and Zhu 1995]. The former gives the connection between possibility measure and evidence theory and the latter gives the connection between possibility and fuzzy membership function. Confidence measure is also known as fuzzy measure or Sugeno measure. Sugeno first proposed this unifying measure [Takagi and Sugeno 1983]. Evidence theory was first proposed by Dempster and Shafer [Shafer 1976] and is also known as upper and lower probability. The advantage based on confidence measure is that the concept of confidence measure can unify the various measures such as the belief measure, plausibility measure, and probability measure and thus gives a theoretical basis of possibility measure. Another reason for the better theoretical basis is because the confidence approach is based on a set of basic axioms. The advantage based on fuzzy set is that the approach offers easy means to obtain numerical values of the representation. In order to save space and also for later applications, only the approach via fuzzy set will be used to define the possibility measure.

The theory of possibility, as the name implies, deals with the possible rather than the probable values of a variable [Zadeh 1987, Lee and Zhu 1995]. The term variable used in here is more in a linguistic sense than a numerical one. Let us consider the statement or proposition, "*X* is *F*", where *F* is a fuzzy set characterized by its membership function μ_F. the variable *X* could be the name of an object, some attribute of an object, a variable, a proposition, or any other similar representations. For example, in "*X* is a small number", *X* is the name of a variable. In "John is tall", John is the name of an object. In "Xelo company's product has very good quality", the word "product" is the name of an attribute of the object "Xelo company". In "The company is very young is quite true", where "The company is very young" is a proposition. The expressions "small number", "tall", "very good quality", and "quite true" are fuzzy sets. Notice that all these expressions are fuzzy or vague and cannot be used to form a crisp set.

4.1.1 Possibility Distribution

One of the central concepts in possibility theory is possibility distribution, which serves the same purpose in possibility theory as probability distribution in probability theory.

In order to define possibility distribution, Zadeh introduced the concept of a fuzzy restriction. We first define a fuzzy concept *F* in the non-fuzzy universe *U*, where the fuzzy concept *F* is characterized by the membership function, $\mu_F(u)$. Then, we consider a given variable *X* and estimate to what degree each of the members of the variable *X* can be assigned to the fuzzy set *F*. Thus, *F* is a fuzzy restriction on the variable *X* if *F* acts as an elastic constraint on the values that may be assigned to *X*. In other words, $\mu_F(u)$ is the degree to which the constraint represented by the fuzzy concept *F* is satisfied when *u* is assigned to *X*. Equivalently, $1 - \mu_F(u)$ is the degree to which the constraint must be stretched in order to allow the assignment of *u* to *X*, where *u* represents the generic elements of *X*.

Let *R*(*X*) denote a fuzzy restriction associated with *X*. Then *R*(*X*) = *F* is called a relational assignment equation by Zadeh, because it assigns the fuzzy set *F* to the fuzzy restriction *R*(*X*).

Let *ACX*) be an attribute of the variable *X*. For example, if *A*(*X*) is the product of Xelo company and F is the fuzzy set "very good quality", the proposition "Xelo company's product has very good quality" can be expressed as:

$$R(A(X)) = F \tag{4.1}$$

In other words, the product of Xelo company is restricted by the fuzzy set "very good quality".

As another example, consider "John is tall". This proposition implied the attribute height of John. Thus, we have:

$$\text{John is tall} \rightarrow R\,(\text{height (John)}) = \text{tall} \tag{4.2}$$

Thus, let F be a fuzzy subset of a universe of discourse U, which is characterized by its membership function μ_F, with the grade of membership $\mu_F(u)$, interpreted as the compatibility of u with the concept label F. Let X be a variable taking on values in U and let F act as a fuzzy restriction, $R(X)$, associated with X. Then, the proposition "X is F", which translates into $R(X) = F$, associates a possibility distribution Π_x with X, which is postulated to be equal to $R(X)$. Thus, the possibility distribution function associated with X, Π_x, can be defined to be numerically equal to the membership function of F, that is,

$$\pi_x \equiv \mu_F \tag{4.3}$$

where "≡" stands for "is defined to be", and π_x represents the possibility distribution function of Π_x.

Therefore, $\mu_F(u)$ not only represents the grade of membership of u in U, but also represents the possibility that X is u given that X is F. Mathematically,

$$\text{Poss}(X \text{ is } u \mid X \text{ is } F) = \mu_F(u),\ u \in U \tag{4.4}$$

Which is a conditional possibility expression parallel to the conditional probability expression.

Example 4.1. Let U be the universe of positive integers and F the fuzzy subset of small positive integers defined by

$$F = \text{small positive integers} = 1.0/1 + 1.0/2 + 0.9/3 + 0.7/4 + 0.5/5 + 0.3/6 + 0.1/7 \tag{4.5}$$

where we have used the conventional method to express the membership function. The plus signs "+" should be interpreted as unions, not as additions and the "/" should not be interpreted as division. The above equation can be interpreted as:

> "The membership function for the integer "1" is 1.0 in the fuzzy set F, the membership function for the integer 3 is 0.9 in the fuzzy set F, the membership function for the integer 5 is 0.5 in the fuzzy set F, and *etc.*"

For example, we can form the following conditional possibility expressions:

$$\text{Poss}(X \text{ is a small integer} \mid X \text{ is } 4) = 0.7 \tag{4.6}$$

$$\text{Poss}(X \text{ is a small integer} \mid X \text{ is } 6) = 0.3 \tag{4.7}$$

4.1.2 Possibility Measure

Based on the above discussion and the equivalence between the membership function of a fuzzy subset and its possibility, the possibility measure can be defined in the following manner. Let A be a non-fuzzy subset of the universe of discourse U and let π_x be the possibility distribution function of Π_x, then, the possibility measure, $\Pi(A)$, of A is defined as a number in the interval [0, 1] and is given by

$$\Pi(A) = \sup_{uA} \pi_x(u) \tag{4.8}$$

where Π_x is the possibility distribution associated with the variable X which takes the value in U. The possibility measure can also be interpreted as the possibility that the value of X belongs to A, or,

$$\text{poss}\ (x \in A) = \Pi(A) = \sup_{uA} \pi_x(u) = \sup_{u \in A} \mu_F(u) \tag{4.9}$$

where the last equality is obtained by using the relationship between the possibility distribution function the membership function of the fuzzy set F.

Example 4.2. Xelo manufacturing company has three divisions. Each division makes a different product for sale. The management estimates that the possibilities of increase 10% in sales for the next fiscal year are 0.7, 0.5, and 0.9 for Divisions one, two, and three, respectively.

Notice that in this example, the universe of discourse is all the divisions which increase sales 10%, and the subset A is either the three given divisions or any combinations of the divisions as long as the individual divisions' increase in sales is 10%. The divisions, which increase sales 10%, are crisp and the subset A is certainly not fuzzy. The fuzzy concept F is the possibility of increasing sales 10%. If we define the crisp subset A as {division 1 increase 10%, division 2 increase 10%, division 3 increase 10%}, then the possibility distribution is

$$\Pi_x = 0.7/(\text{division 1 increase 10\%}) + 0.5/(\text{division 2 increase 10\%}) + 0.9/(\text{division 3 increase 10\%}) \tag{4.10}$$

According to Equation 4.8 or Equation 4.9, the possibility of increasing 10% for divisions 1, 2, and 3 as one entity is

$$\Pi(A) = \Pi(\{\text{division 1 increase 10\%, division 2 increase 10\%, division 3 increase 10\%}\}) = \sup\ \{0.7, 0.5, 0.9\} = 0.9 \tag{4.11}$$

If we define subset A as composed of only two divisions, then the possibility of increasing 10% is:

$\Pi(A) = \Pi(\{\text{division1 increase 10\%, division 2 increase 10\%}\}) = \sup\{0.7, 0.5\} = 0.7$ (4.12)

Notice that the above definition of possibility is consistent with the meaning of the word "possibility" in our usual daily usage.

Example 4.3. Suppose Xelo Company is starting to produce a new product. From the various economical data about the product, the company concluded that if the total sale during the first year is about 35 million pounds, then this product would produce a profit. But, many economic and other factors can influence this estimate. Based on past experience, plus or minus 5 million pounds should be allowed for this rough estimation. Thus, the membership function for the profit potential or making a profit can be represented as:

$$\mu_{\text{profit potential}}(u) = \begin{cases} 1, & 40 \le u \\ (u-30)/10, & 30 \le u \le 40 \\ 0, & u \le 30 \end{cases} \tag{4.13}$$

Suppose the total sales of this new product during the first year is 35 million pounds, then the profit potential or the possibility of making a profit during the first year is:

$\text{Poss (profit} \mid \text{amount produced} = 35 \text{ million pounds)} = \text{poss (making a profit)} = \mu_{\text{profit potential}}(35) = 0.5$ (4.14)

4.1.3 Possibility Measure Based on Fuzzy Set with Fuzzy Subset

In the above discussions, the possibility measure was defined on a crisp or non-fuzzy subset A. When A is a fuzzy subset, the belonging of a value of X to A has no meaning. A more general definition for possibility measure with fuzzy subset A was proposed by Zadeh and can be introduced as follows.

Let A be a fuzzy subset of the universe U and let Π_x be a possibility distribution associated with a variable X which takes value in U. The possibility measure, $\Pi(A)$, of A is then defined by

$$\text{Poss}\{X \text{ is } A\} = \Pi(A) = \sup_{u \in U} \min[\mu_A(u), \pi_x(u)] \tag{4.15}$$

where $\mu_A(u)$ is the membership of u in the fuzzy subset A. The above equation can be interpreted as the possibility that X takes fuzzy set A as its value.

Example 4.4. Consider the problem in Example 4.3. In addition to the economic considerations of the new product, the sales or demand potential of the product must also be considered. From marketing survey, it was found out that the demand for this product during the first year is about 35 million pounds. Because various factors can influence this survey, plus or minus 5 million pounds should be allowed to this figure. Thus, the membership function for demand potential can be expressed as (see Figure 4.1):

$$\mu_{\text{demand potential}}(u) = \begin{cases} 1, u \leq 30 \\ (40 - u)/10, 30 \leq u \leq 40 \\ 0, u \geq 40 \end{cases} \tag{4.16}$$

Using the membership functions of profit potential, Equation 4.8, and demand potential, Equation 4.16, The possibility of making a profit based on Equation 4.15 is:

$$\text{Poss(making a profit)} = \sup \min (\mu_A(u), \mu_X(u)) = \sup \min (\mu_{\text{sales potential}}(u), \mu_{\text{profit potential}}(u)) = 0.5 \tag{4.17}$$

Example 4.5. Consider the problem of Xelo Company again. Although the new plant for the new product was built for a capacity of 35 million pounds, the actual on-line capacity may be different due to many factors, which cannot be controlled. From past experience, the actual capacity may be anywhere within the interval of [30 million pounds, 40 million pounds]. It should be noted that because of the particular process, it is very uneconomical, in fact, may be dangerous to produce too much below capacity. Thus, the membership function for capacity should reach a value of one at 35 million pounds, or this is the most possible value. Assuming the possibility decreases linearly when the capacity moves away from 35 from either side and reaches a value of zero at either 30 or 40 million pounds. Thus, the membership function for the approximate capacity of production is (see Figure (4.1):

$$\mu_{\text{capacity}}(u) = \begin{cases} 0, u \leq 30 \text{ or } u \geq 40 \\ (u - 30)/5, 30 \leq u \leq 35 \\ (40 - u)/5, 35 \leq u \leq 40 \end{cases} \tag{4.18}$$

Assume that the company has the least confidence about the profit potential and the estimated approximate capacity, then an estimate on possibility of making a profit based on these two fuzzy numbers can be obtained by using Equation 4.15. The result is:

$$\text{Poss (making a profit)} = \sup \min (\mu_A(u), \mu_X(u)) = \sup \min (\mu_{\text{profit potential}}(u), \mu_{\text{approximate capacity}}(u)) = 0.67 \tag{4.19}$$

If, on the other hand, the company is worried more about capacity and demand potential, the possibility of making a profit is:

$$\text{Poss (making a profit)} = \sup \min (\mu_A(u), \mu_X(u)) = \sup \min (\mu_{\text{approximate capacity}}(u), \mu_{\text{demand potential}}(u) = 0.67 \quad (4.20)$$

Finally, if we take all three factors into consideration, that is profit potential, market demand and the approximate capacity, then the possibility of making a profit is:

$$\text{Poss (making a profit)} = \sup \min (\mu_{\text{profit potential}}(u), \mu_{\text{demand potential}}(u), \mu_{\text{approximate capacity}}(u)) = 0.5 \quad (4.21)$$

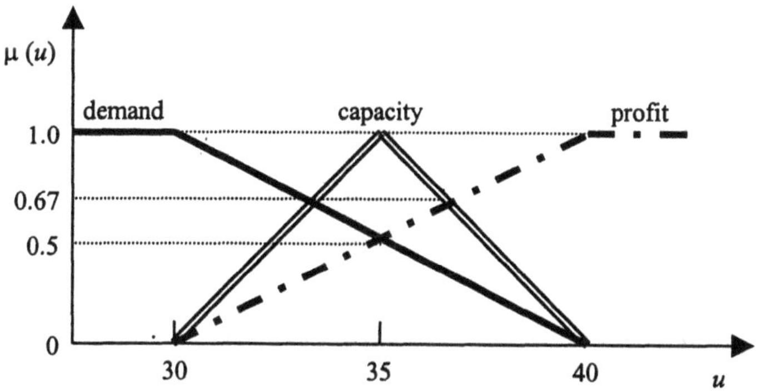

Figure 4.1. Membership functions of Xelo Company

4.1.4 Possibility versus Probability

Possibility deals with the possible while probability deals with the probable. Possibility is associated with the degree of feasibility or the sense of attainment, whereas, probability relates to the degree of likelihood, frequency, or proportion. This implies that what is possible may not be probable, and what is improbable may not be impossible.

There are many types of uncertainties. Probability models randomness while possibility models fuzziness where the events have no clear boundary. The concept of possibility, unlike probability, does not involve the notion of repeated experimentation. Thus, possibility can be used to model the imprecision of uncertainties, which are not susceptible to probability analysis or characterization [Lee and Zhu 1995].

Another way to compare these two different approaches of modeling uncertainties is to consider the rules, which govern their aggregation. Possibility theory based on the max-min aggregation while probability theory requires that the summation of the probabilities of all the possible outcomes must be equal to one. In possibility, intersection is modeled by the min-operator and union is modeled by the max-operator, while in probability, intersection of two independent events corresponds to the product operator and the union of two independent, mutually exclusive events corresponds to the sum operator. Thus, in a sense, the requirements of possibility are not as strict as that of probability. In fact, it has been proved that possibility measure reduces to probability measure if only singletons are used in assigning the amount of believe.

Many of the particular characteristics of possibility theory are intimately connected with linguistics. For example, human language is frequently vague and approximate, which is exactly the same for fuzzy or possibility. Language is intimately connected with human activities and thus it is subjective and cannot be objective, fuzzy set is also subjective, not like the objective requirements when probability is used. One example would clear the reason why language or fuzzy set must be subjective. Consider the word "young' in the phrase "a young man". Since it is intimately connected with our concept, which represents "young", it would be impossible to define the word "young" objectively.

Perhaps the most important difference is that fuzzy set is best for describing humanistic aspects and probability is more for mechanistic or scientific applications.

4.2 Knowledge Representation

Possibility theory based on fuzzy membership function or fuzzy logic forms a useful approach for knowledge representation and subsequent manipulation or aggregation on computer. It is much more flexible and capable of linguistic representation than traditional logical systems. In other words, the traditional proposition logic is too restrictive for linguistic representation and aggregation. Some of the problems with the traditional two-value logic are:

1. in the traditional logic, a proposition is either true or false. But, human language is generally vague or fuzzy and it seldom only has the two extremes;
2. the predicates in two-value logic are constrained to be crisp. But, human language frequently contains fuzzy or vague predicates;
3. traditional logic only allows two quantifiers: *all* and *some*. By contrast, human language uses many quantifiers, both fuzzy and crisp, such as *many, much, most, few, frequently, etc.*

Since linguistics represents human intelligence and human knowledge, knowledge or linguistic representation in computer constitutes the essential ingredient in the use of computer for solving and manipulating soft science such as management or decision making, which is frequently expressed in linguistic terms.

One of the most difficult problems encountered is the representation of common sense knowledge, some typical examples of which are:

birds fly;
snow is white;
this manufacturing process is reliable.

The first two propositions were well known and were first encountered in the investigation of expert systems. One characteristic of these propositions is that it is preponderantly, but not always, true. This type of daily encountered common sense knowledge is the most difficult problem to handle. However, they can be handled elegantly by the use of fuzzy concepts. Notice that these propositions have implicit or implied fuzzy quantifiers. For example, *birds fly* should be *most birds fly* and *snow is white* should be *snow is usually white.*

In this section, knowledge representation will be treated in terms of linguistic representation. Linguistic variable and linguistic terms will be first defined. The syntactic and semantic rules of the linguistic variable will then be discussed. And, finally, linguistic algorithm for the manipulation of the representations will be presented. Because of space limitations, only a summary can be presented. The interested reader can consult the literature [Lee and Zhu 1992, 1995, Zadeh 1987].

4.2.1 Linguistic Variable[1]

Linguistic variable forms the first step or basis in the manipulation or aggregation of human language on computer, or, computing with words, as Zadeh proposed. Since human intelligence is embodied in the human language, in order to manipulate or aggregate human intelligence, it is essential to be able to aggregate the human language on modern computer.

Like the traditional or numerical variable, linguistic variable also has values. For example, consider the linguistic variable, *price to earnings ratio (PE ratio),* which is usually used to describe the expensiveness of the stock of a company. The values of this linguistic variable may be "the stock of the company is *very expensive, fairly expensive, expensive, average, cheap, fairly cheap, very cheap, etc.*" It should be noted that the expensiveness of a company's stock is, in general, a very vague expression and cannot depend on its numerical value. A very high PE ratio for a very fast growing company is not necessarily expensive, and, on the other hand, a very low PE ratio for a company, who is continuously loosing its market share, may be still too expensive. Thus, instead of using numerical values to describe the expensiveness of a company's stock, it is much more reasonable to use the above mentioned linguistic terms.

Zadeh [1987] characterized a linguistic variable by the quintuple $(L, T(L), U, S, M)$ in which L is the name of the variable such as the linguistic variable *PE ratio*; $T(L)$ is the term-set of L or the collection of the linguistic values such as *very expensive, fairly expensive, average, etc.*, in the above example; U is the universe of discourse; S is the syntactic rule, by the use of which the term-set X for L can be generated; and M is the semantic rule which associates with each linguistic value X its meaning $M(X)$ by the use of the possibility concept and fuzzy set. Notice that $M(X)$ is a fuzzy subset of U and the concept of linguistic variable is defined based on possibility and the fuzzy concept.

A linguistic variable can generally be represented by a three- or two-level hierarchy system. This is shown in Figure 4.2, where the linguistic variable *Age*, as used by Zadeh, is reproduced [1978, 1987]. The lowest hierarchy is the base

[1] [Lee and Zhu, Zadeh]

variable, which usually is a subset of the universe of discourse. In the present case, the base variable is the actual age. The middle hierarchy is the term-set, or the linguistic value of the variable. The terms of the term-set for *Age* are *very young, fairly young, young,, middle age, old,, fairly old,, very old, etc*. The membership functions for the term-set are also shown in the figure.

For some linguistic variables, there exist no base variables and thus a two-level hierarchy must be used. One example is the linguistic variable describing the ability of the manager of a company, *Manager*, whose term-set may be *very good manager, good manager, average manager, bad manager, very bad manager, etc*. Another example is the linguistic variable *appearance*, whose term-set may be *very neat, fairly neat, neat, average, sloppy, fairly sloppy, very sloppy, etc*. There exist no base variables for these linguistic variables. In general, more precise and meaningful representation, in terms of membership function, can be obtained if the base variable for this linguistic variable can be established.

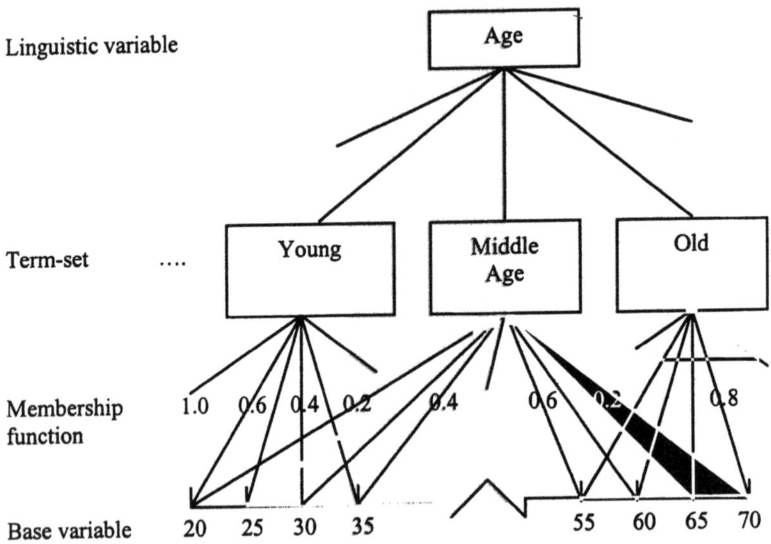

Figure 4.2. The linguistic variable "Age" [Zadeh 1975]

4.2.2 The Syntactic Rule

According to Zadeh's approach, there is no limit on the number of terms or elements in a term-set. However, there exists a definite structure for the term-set of any linguistic variable. A term-set consists of primary terms, which are usually finite in number, and composite terms, which are various combinations or modifications of the primary terms. For example, for the linguistic variable *PE* ratio, the primary terms are *expensive* and *cheap*. An infinite number of composite terms can be obtained such as *very expensive, fairly expensive, more or less expensive, not expensive not cheap, cheap, more or less cheap, fairly cheap, not quite cheap, extremely cheap, etc.*

Most linguistic variables have the same basic structure of term-set. For example, one can replace *cheap* by *young* and *expensive* by *old*, one obtains the primary terms for the linguistic variable *Age*. The same applies to many other linguistic variables such as the goodness of a *management* (*good* and *bad*), the rate of increase of a company's *profit* (*high* and *low*), *distance* (*far* and *near*), *truth* (*true* and *false*}, *quality* (*good quality* and *bad quality*), *weight* (*heavy* and *light*), *etc.*

As has been pointed out by Zadeh, the composite term is a concatenation of atomic terms. The atomic terms generally used can be divided into the following four categories:

1. primary terms;
2. the negation *not* and the connectives *and* and *or*;
3. linguistic hedges such as *very, fairly*, extremely, *etc.*;
4. markers such as parentheses.

Some linguistic hedges are composed more than one word such as *more or less*, which is still considered as an atomic term.

The approach is based on Boolean operations and thus Zadeh called the variable the Boolean linguistic variable. Based on the four categories of atomic terms, the Boolean linguistic variable usually consists of primary terms, hedges, the connectives *and* and *or,* and the negation *not*. For example, the term-set of the linguistic variable *PE ratio* may be:

$$T(\text{PE ratio}) = \text{expensive} + \text{cheap} + \text{not expensive} + \text{not cheap} + \text{very expensive} + \text{very cheap} + \text{more or less expensive} + \ldots \tag{4.22}$$

where the "+" sign should be interpreted as union.

The Boolean linguistic variable can be defined recursively. Let x_p represents the primary term and h a linguistic hedge, then the composite term can be formed as hx_p, which is a fuzzy set resulting from acting with h on x_p. To continue this operation and upon further acting by the linguistic hedge h on hx_p, we obtain hhx_p. For example, if the hedge is *very* and the primary term is *expensive*, then we first have *very expensive*. Upon further acting, we obtain *very very expensive*.

As another example, consider the linguistic variable *truth*, whose truth value in classic logic can only be either true or false. However, in fuzzy logic, *true* and *false* form the primary terms. We now have various degree of *true* and *false*, which can be formed by the use of various hedges to form composite terms. Thus, the term-set for the linguistic variable *truth* may be *very true, fairly true, more or less true, not true, not false, very false, fairly false, etc.*

4.2.3 The Semantic Rule

The meaning of the Boolean linguistic variable can also be obtained based on the above mentioned four categories of atomic terms. The two categories need further discussions are the meanings of the Boolean negation *not* and connectives *and* and *or* in the fuzzy usage and the linguistic hedges. According to Zadeh, the linguistic hedges can approximately be divided into the following two types:

Type 1. Hedges which can be represented as operators acting on fuzzy set such as *very, plus, minus, highly, much, more or less, etc.*

Type 2. Hedges which require a description of how they act on the components of the operand such as *essentially, technically, actually, strictly, practically, regularly, virtually, etc.*

The first type is easy to interpret. For example, the hedge *very* can be considered as *concentration*, which serves as intensifier. Thus,

$$\text{Very good manager} = (\text{good manager})^2 \tag{4.23}$$

which can be generalized as:

$$\text{Very } x_p = \text{con}\,(x_p) = (x_p)^2 \tag{4.24}$$

The phase *more or less* can be interpreted in different ways. One way is to interpret this hedge as dilation

$$\text{More or less } x_p = (x_p)^{0.5} \tag{4.25}$$

Type 2 hedges are much more complicated because they affect the components of the operand. For example, the hedge *essentially* has the effect of increasing the important attributes and diminishing those that are relatively unimportant. Consider the phrase *John is essentially a good manager*, where *good manager* might be composed of the attributes: *smart* (S), *decisive* (D), *good leadership ability* (L), *fair* (F), *etc.* The effect *essentially* would tend to increase the weights of the important attributes in the following convex combination:

$$\mu_{\text{good manager}}(\text{John}) = w_1\mu_S(\text{John}) + w_2\mu_D(\text{John}) + w_3\mu_L(\text{John}) + w_4\mu_F(\text{John}) \tag{4.26}$$

where the w's are the weights. To be nonbiased, the summation of all the w's must be equal to one. Which one or ones are the important attributes must be decided by the decision-maker.

Thus, if the membership functions of the operands can be expressed as a convex combination, then the effect of the hedges such as *essentially, technically, regularly, practically, etc.*, is to increase the magnitudes of the weights more for those attributes, which have relatively high membership function values. However, if the membership function cannot be expressed as convex combination, then the characterization of the Type 2 hedges become more complicated and may require the solution of fuzzy algorithms.

Using the linguistic variable and term set concepts, complicated and vague linguistic expressions such as recession, inflation, intelligence, creativity, *etc.*, can be formulated as fuzzy algorithms and can be represented in modern computers. The following simple example due to Zadeh illustrates the essence of the approach.

Example 4.6. The typical characteristics of a *hippie* can be described as: A person who is unemployed, uses drugs and has long hair, provided that the person is not bald. The fuzzy algorithm becomes:

$$\text{Hippie} = (\text{drug}) \cap (\neg(\text{employed})) \cap (\text{long hair} \cup \text{bald}) \tag{4.27}$$

In terms of membership functions for hippie John, we have:

$$\mu_{\text{hippie}}(\text{John}) = \mu_{\text{drug}}(\text{John}) \wedge (1 - \mu_{\text{employed}}(\text{John})) \wedge (\mu_{\text{long hair}}(\text{John}) \vee \mu_{\text{bald}}(\text{John})) \tag{4.28}$$

The interpretation of Boolean negation and connectives, or, more generally, the fuzzy logic connectives such as negation $\neg$, conjunction $\wedge$, disjunction $\vee$ and implication $\Rightarrow$, are, more complicated. This is especially true for fuzzy implication. There are many different interpretations. One approach is to base the interpretation on Zadeh' extension principle. Although the extension principle approach is theoretically sound, they are not easy to use due to computational complexity. A simpler approach is to define a simpler fuzzy truth space and interpret the connectives based on this simpler definition [Baldwin 1979]. The reader is referred to the literature [Lee and Zhu, Baldwin] for more detail.

4.2.4 Test-score Semantics

Zadeh proposed the concept of test-score semantics to provide a framework for the representations of the meanings of various types of semantic entities such as propositions, predicates, dispositions, and commends. The idea underlying the concept is that a proposition in a natural language may be interpreted as a system of fuzzy restrictions or constraints, which are analogous to a non-linear programming problem. Thus, the problem can be handled in the same general fashion as in non-linear programming. However, since we are dealing with linguistic quantities, not numbers, the concept of possibility and fuzzy set must form the basic approach, and, at the same time, the use of modern computer to express linguistic quantities can be achieved. The proposed approach scores the concerned proposition by first identify the elastic constraints; then by the use of an explanatory database (ED) to score the collection of fuzzy relations; and, finally aggregate the various scores to arrive at a composite single or vector score. The ED forms the basic reference or standard for establishing the fuzzy membership functions. A linguistic proposition is viewed as composed of various fuzzy relations whose meaning is assumed known to the user. As proposed by Zadeh, the procedure of representing the meaning of a proposition p, through the use of test-score semantics involves the following steps [Zadeh 1987, Zhu and Lee 1995]:

1. *Identification of Variables.* The variables $X_1, X_2, \ldots, X_n$, whose values are constrained by the proposition must be first identified. Usually these variables are implicit in p. For example, if the proposition is p = John is a good manager, the implicitly constrained variables are the various attributes of John, which make him a good manager.
2. *Identification of Constraints.* The constraints $C_1, C_2, \ldots, C_m$, which are induced by p must be identified. Sometimes the constraints are implicit also. This is especially true for dispositions. For example p= It takes 5 hours to fly from New York to London. The constraint should be *usually, or,* "*Usually* it takes 5 hours to fly from New York to London".
3. *Characterization of Constraints.* Each constraint C_i must be characterized or represented by a test procedure, which associates with C_i a test score τ_i representing the degree to which C_i is satisfied. Usually, τ_i is a number in the interval [0, 1]. More generally, τ_i may be a probability or possibility distribution over the unit interval.
4. *Aggregation of Test-scores.* the various test-scores $\tau_1, \tau_2, \ldots, \tau_m$ must be aggregated into a smaller number of test scores such as $\tau_1, \tau_2, \ldots, \tau_k$, which are represented as an overall vector test score. In most cases, $k = 1$.

To illustrate the general concept, consider the simple example used by Lee and Zhu [1995]:

Example 4.7. Given the fact that John is 20 years old and 6 feet tall, we wish to consider the proposition:

$$P = \text{John is young } (Y) \text{ and tall } (T) \tag{4.29}$$

Let

$$\text{Young} = \mu_Y(X_1),\ \text{Tall} = \mu_T(X_2)$$

1. *Identification of Variables.* There are two variables in p, which are:

$$X_1 = \text{Age},\ X_2 = \text{height}$$

2. *Identification of Constraints.* The constraints are:

$$C_1 = \text{young},\ C_2 = \text{Tall}$$

3. *Characterization of Constraints.* Now, the constraints must be characterized by the use of their respective test scores under proposition p. This is essentially the formulation of the membership functions, which can represent the elastic constraints just identified in (2). This membership function can be represented as:

$$\tau_1 = \mu_y(20),\ \tau_2 = \mu_T(6)$$

4. *Aggregation of Test-scores.* The two partial scores can be aggregated into one score:

$$\tau = \tau_1 \wedge \tau_2 = \min(\tau_1, \tau_2)$$

In the above example, both variables are implicit and all the test scores are scalars. If the scores are possibility distributions, Zadeh's extension principle can be used to aggregate the partial scores. For example, consider the linguistic variable *good manager* defined in Equation 4.26 and the proposition:

$$P = \text{John is a good manager} \tag{4.30}$$

Among the four attributes of a good manager, may be only *smart (S)* has a base variable, which can be represented by an IQ from 100, which is 50 percentile, to 150. Even though, theoretically, the upper limit of IQ can be infinity. Noticed that, again for simplicity, we have defined the boundary between *smart* and *not smart* as a crisp separation and not fuzzy. The other attributes have no base variable. For example, the attribute *good leadership ability* has no base variable and is usually represented in linguistic terms such as: *very low L, low L, average, high L, and very high L. Decisive* and *fair* can be described by the same set of terms. Each linguistic term can be represented by a triangular membership function. Notice that the variable for *smart* is implicit and IQ can be specified with a crisp number. However, other attributes must be represented by the use of fuzzy numbers.

Chapter 5

Fuzzy Decision Making

After formulating the fuzzy representations of the various linguistic terms, the next step is decision-making based on these formulations and by using fuzzy optimization or decision making techniques. Fuzzy optimization approaches have been developed for most of the existing crisp optimization and decision making techniques. Fuzzy linear programming is a well-developed field and has been extended to fuzzy multiple-objective programming and fuzzy goal programming. Fuzzy multiple attribute decision making is an ideal approach to overcome the representation of the frequently encountered linguistic expressions such as *very good, good, fairly good, average, fairly poor, poor, and very poor.* For dynamic and non-linear systems, there are fuzzy dynamic programming and various fuzzy non-linear programming techniques. Many of these techniques can generally be divided into two types. One is based on the original crisp approach with the addition of fuzzy aspects and the other handles the fuzzy problem directly, which frequently forms completely new approaches. One good example is the newly formed possibility programming.

Because of the vast development in this area, it is clear impossible, nor desirable, to have even a simple comprehensive review. Only some basic concepts and the ones, which will be used in latter chapters, will be summarized. The interested readers can consult the references listed at the end of the book [Zimmermann 1996, Chang and Lee 1993].

5.1 Fuzzy Linear Programming

The very first contribution in fuzzy decision making is due to Bellman and Zadeh [1970], who proposed the concept of symmetry between the objective function or goal and the constraint. Many later developments are based on this concept. These investigators argued that for complicated problems in actual situations, it is more practical to consider the objective is to achieve a desired fuzzy goal, which can be treated symmetrically with the constraint. In order to simultaneously satisfy all the constraint and the goal, the minimum or the conjunction *and* should be used. Thus, we can formulate the following:

$$\mu_D = \mu_G \wedge \mu_C = \min(\mu_G, \mu_C) \tag{5.1}$$

where μ_D, μ_G and μ_C represent respectively the membership functions of decision, goal and constraint. In order to obtain the maximum value of the membership function of the final decision, the above expression is maximized:

$$(\mu_D)_{\max} = \max(\min(\mu_G, \mu_C)) \tag{5.2}$$

Based on this symmetric concept, Zimmermann [1978, 1996] formulated fuzzy linear programming in the following manner. Consider the typical linear programming problem:

$$\max\ z = c^T x \tag{5.3}$$

subject to:

$$Ax \le b$$
$$x \ge 0$$

where x and c are n-dimensional vectors, b is m-dimensional and A is an $m \times n$ matrix. In many practical applications and especially for large practical systems, we frequently prefer to achieve an established goal instead of maximizing the objective function. Assuming the established aspiration level or goal as z, Equation 5.3 reduces to:

$$\begin{aligned} c^T x &\le z \\ Ax &\le b \\ x &\ge 0 \end{aligned} \tag{5.4}$$

which can be stated as finding the values of x such that:

$$\begin{aligned} Bx &\le d \\ x &\ge 0 \end{aligned} \tag{5.5}$$

where the first two parts of Equation 5.4 have combined to form the first part of Equation 5.5. The first vector equation in Equation 5.5 represents $(m + 1)$ equations. In order to satisfy all the inequality constraints in the first part of Equation 5.5, the membership function of the decision must obey the following equation:

$$\mu_D(x) = \min_i(\mu_i(x)) \tag{5.6}$$

where the minimization is with respect to all the equations in the first vector equation in Equation 5.5. The results of Equation 5.6 are represented by a fuzzy set. If the decision-maker is interested in a crisp optimum, we should take the maximum possible value of the membership function. Thus, we have:

$$\mu(x) = \max_{x \geq 0} (\min_i (\mu_i(x)) \tag{5.7}$$

which can also be written as:

$$\mu(x) = \max_{x \geq 0} \lambda \tag{5.8}$$

with

$$\lambda = \min_i (\mu_i(x)) \tag{5.9}$$

In practice, frequently, the inequality constraints in Equation 5.3 are not strictly obeyed. Thus, we can also assume the inequalities are fuzzy and the fuzzy membership functions are:

$$\mu_i(x) = \begin{cases} 1, \text{ if } (Bx)_i \leq d_i \\ 1 - \dfrac{(Bx)_i - d_i}{p_i}, \text{ if } d_i \leq (Bx)_i \leq d_i + p_i \\ 1, \text{ if } (Bx)_i \geq d_i + p_i \end{cases} \tag{5.10}$$

where $i = 1, 2, \ldots, m+1$ and p_i is the allowed violations of the objective and constraints or the allowed violations of constraint i in the first part of Equation 5.5. This allowed violation must be determined by the decision-maker. Substituting Equation 5.10 into Equation 5.9, we have:

$$\lambda = \min_i (1 - ((Bx)_i - d_i)/p_i) \tag{5.11}$$

The minimization sign can be replaced by inequality, we have:

$$\lambda \leq (1 - ((Bx)_i - d_i)/p_i) \tag{5.12}$$

after simplification of which, we have:

$$\lambda\, p_i + (Bx)_i \leq d_i + p_i, \quad i = 1, 2, \ldots, m+1 \tag{5.13}$$

Combine with Equation 5.8, we obtain the new linear programming problem:

$$\max \lambda \tag{5.14}$$

subject to:

$$\lambda p_i + (Bx)_i \leq d_i + p_i, \quad i = 1, 2, \ldots, m+1$$
$$x_j \geq 0, \quad j = 1, 2, \ldots, n$$

In general, the solution of Equation 5.14 gives the solution of the original fuzzified linear programming problem. By fuzzified, we mean that we allowed some tolerance to the original strict inequalities and used a predefined goal to replace the maximization operation. Computationally, Equation 5.14 only increased one constraint as compared to the original linear problem, Equation 5.3. Thus, this approach is used frequently due to its ease in obtaining numerical solution. Many extensions in fuzzy decision-making such as fuzzy multi-objective programming are also frequently based on this approach.

5.2 Multiple-objective Programming

One of the difficulties in multi-objective decision making is the aggregation or compromise between the various objectives. Various approaches to carry out this aggregation have been proposed and they can be classified approximately into compensatory and non-compensatory aggregation operators. Zimmermann and coworkers [1978, 1996] have proposed a compensatory *and* γ operator, which was a combination of the algebraic product and the algebraic sum. However, due to the ease of computation, the most frequently used aggregate operator is the *min* operator, which is completely non-compensatory. To overcome this non-compensatory problem, Lee and coworkers [1993, Li 1990] proposed a two phase approach, where the non-compensatory operator *min* was used in the first phase to obtain the optimal degree of the overall satisfaction. Then a fully compensatory operator *averaging* was introduced in the second phase to find a non-dominated solution. It was shown that, in this way, a non-dominated and balanced solution could be obtained in phase two, regardless of the uniqueness of the solution.

The general multiple-objective programming problem can be represented as:

$$\max Z = [Z_1, Z_2, \ldots, Z_\ell]^T = [c_1x, c_2x, \ldots, c_kx, \ldots, c_\ell x]^T$$
$$\min W = [W_1, W_2, \ldots, W_r]^T = [c_1x, c_2x, \ldots, c_sx, \ldots, c_rx]^T \qquad (5.15)$$

subject to:

$$Ax \leq b$$
$$x \geq 0$$

where c_k, c_s, and x are n-dimensional vectors with $k = 1, 2, \ldots, \ell$ and $s = 1, 2, \ldots r$, b is an m-dimensional vector, and A is an $m \times n$ matrix. Many different approaches have been proposed to solve this problem. Only two of the more frequently used approaches, namely, compromise programming and goal programming, will be summarized below.

5.2.1 Compromise Programming

The compromise programming approach is to obtain the ideal and anti-ideal solutions first. The ideal solution can be obtained by solving each objective function independently subject to the constraints in Equation 5.15. This ideal solution can be represented by:

$$I^* = (Z_1^*, Z_2^*, \ldots, Z_\ell^*; W_1^*, W_2^*, \ldots, W_r^*) \tag{5.16}$$

The anti-ideal solution of Equation 5.15 can be obtained by minimizing the profit objective functions, Z, and by maximizing the cost functions, W. All these minimization and maximization operations are carried out with each objective function independently and subject to the constraints in Equation 5.15. The anti-ideal solution can be represented by:

$$I^- = (Z_1^-, Z_2^-, \ldots, Z_\ell^-; W_1^-, W_2^-, \ldots, W_r^-) \tag{5.17}$$

To obtain a compromise solution, the following distance expression is minimized [Li 1990]:

$$D_p = \sum_{k=1}^{\ell} \alpha_k{}^p \left[\frac{Z_k^* - Z_k(x)}{Z_k^* - Z_k^-}\right]^p + \sum_{s=1}^{r} \alpha_s{}^p \left[\frac{W_s(x) - W_s^*}{W_s^- - W_s^*}\right]^p \tag{5.18}$$

where p is a distance parameter with values $1 \le p \le \infty$ and α_k and α_s represent the relative importance or weights of each of the objective functions with $\alpha_k \ge 0$, $\alpha_s \ge 0$. The summation of all the weights must be equal to one. The frequently used values for p are 1, which is known as the city block or Manhattan distance, 2, which is the Euclidean distance, and ∞, which is the Tchebycheff distance. D_1 and D_2 are respectively corresponding to the longest and the shortest distance between two points in Euclidean space.

Observing the effect of increasing p on the relative contributions of the individual deviations, the larger the p, the greater the contribution of the largest deviation. If p increases to infinity, then the largest deviation completely dominates the value of D_p. Thus, as p approaches infinity, Equation 5.18 becomes:

$$D_\infty = \max_{k,s} \left[\alpha_k \frac{Z_k^* - Z_k(x)}{Z_k^* - Z_k^-}, \alpha_s \frac{W_s(x) - W_s^*}{W_s^- - W_s^*}\right] \tag{5.19}$$

The maximization in the above equation can be replaced by inequality. Also, since we wish to obtain the minimum distance between the ideal and the non-ideal solutions, the above problem is reduced to the following compromise programming problem:

$$\min D_\infty \tag{5.20}$$

subject to:

$$D_\infty \geq \alpha_k \frac{Z_k^* - Z_k(x)}{Z_k^* - Z_k^-}, \quad k = 1,2,\ldots,\ell$$

$$D_\infty \geq \alpha_s \frac{W_s(x) - W_s^*}{W_s^- - W_s^*}, \quad s = 1,2,\ldots,r$$

$$Ax \leq b$$

$$x \geq 0$$

To illustrate the approach, consider the following problem due to Zimmermann [1978]:

Example 5.1. One export-oriented country is concerned with the import export balance of the products of the ABC Company, which produces two products. Product 1 yields a profit of \$1 per unit and product 2 \$2 per unit. Product 1 can be exported, yielding \$2 per unit and product 2 needs import raw materials at \$1 per unit produced. Let x_1 and x_2 represent the total units produced for product 1 and 2, respectively, then, the problem can be formulated as:

$\max f_1 = 2x_1 - x_2$ import export balance
$\max f_2 = x_1 + 2x_2$ profit

subject to:

$3x_1 - 5x_2 \leq 15$ capacity
$3x_1 - x_2 \leq 21$ management
$3x_1 + x_2 \leq 27$ space
$3x_1 + 4x_2 \leq 45$ material
$x_1 + 3x_2 \leq 30$ labor
$x_1, x_2 \geq 0$

This problem is solved by the compromise programming algorithm. The ideal solution for this problem is $f_1(x_1, x_2) = 13.5$ and $f_2(x_1, x_2) = 21$, and the anti-ideal solution $f_1(x_1, x_2) = -10$ and $f_2(x_1, x_2) = 0$. Substituting these ideal and anti-ideal solutions into the above equations and with the parameter $p = 1$, or the Manhattan distance, a linear programming problem was obtained, which can be solved easily. The solution is $f = (f_1, f_2) = (8, 19)$ with $x = (x_1, x_2) = (7, 6)$.

With the parameter $p = 2$, or the Euclidean distance, a problem with non-linear objective function was obtained. The solution is $f = (9.33, 17.67)$ with $x = (7.27, 5.2)$.

With $p = \infty$, or the Tchebycheff distance, Equation 5.20 was solved and the solution is $f = (9.54, 17.47)$ with $x = (7.31, 5.08)$. Although three different solutions were obtained by the use of three different distance definitions, all these three solutions are located on the non-dominated boundary.

5.2.2 Goal Programming

Goal programming [Charnes and Cooper 1977] is probably the oldest approach for solving the multi-criteria decision-making problem. It is an optimization technique designed to handle decision-making situations where a number of conditions characterized as goals are to be met as closely as possible. A general linear goal programming with K goals can be expressed as:

$$\min \sum_{r=1}^{K} w_r \, a_r(d) \tag{5.21}$$

subject to:

$$x \in X$$
$$g_j(x) + d_j^- - d_j^+ = b_j, \forall j$$
$$g_i(x) - d_i^- + d_i^+ = b_i, \forall i$$
$$g_p(x) - d_p^- + d_p^+ = b_p, \forall p$$

where

$g_j(x)$: benefit criterion for the goal
$g_i(x)$: cost criterion for the goal
$g_p(x)$: two-sided goal
$a_r(d)$: achievement function, which is defined as:

$$a_r(d) = d_r^-, \forall j$$
$$a_r(d) = d_r^+, \forall i$$
$$a_r(d) = d_r^- + d_r^+, \forall p$$

where $j + i + p = k$; w_r represent the relative importance of the different goals, which can be pre-emptive, non-preemptive, or mixed weight and X represents the constraint set in Equation 5.15. Notice that g_k instead function f_k was used in the above equation. This is because we need to distinguish between the objective, which was used in compromise programming, and the goal in goal programming.

A case similar to the multi-level system is the pre-emptive goal programming, which describes the level in a hierarchy as a pre-emptive sequence from top to bottom. The weight of the goal of the higher level dominates that of the lower level, *i.e.*, $w_r^+ \gg w_{r+1}^+$ (or $w_r^- \gg w_{r+1}^-$), where assuming level r is higher than level $r + 1$. Therefore, the first-level goal should be met first, then the second level, and so on. Whitford and Davis [1983] used this goal programming concept to solve a resource allocation problem for multi-level organizations.

Example 5.2. Consider again the export-import problem of Example 5.1. The two objectives are in pre-emptive order, or, the priority of objective 1 is higher than that of objective 2. The solution of the problem by using the pre-emptive goal programming algorithm is $x = (x_1, x_2) = (7.5, 1.5)$ with $f = (f_1, f_2) = (10.5, 13.5)$, which is actually objective 1's problem since it dominates the second objective.

5.3 Fuzzy Approach to Multiple-objective Programming

Fuzzy concepts can also be used to obtain the compromise of the various objective functions of the above crisp programming problem. In fact, it has been shown that the crisp compromise programming approach with $p \Rightarrow \infty$ is equivalent to the following fuzzy approach [Lee and Li 1995]. The first step is to formulate the membership functions for the objectives in the following manner:

$$\mu_k = \frac{Z_k^* - Z_k(x)}{Z_k^* - Z_k^-}, \quad k = 1,2,\ldots,\ell \tag{5.22}$$

$$\mu_s = \frac{W_s(x) - W_s^*}{W_s^- - W_s^*}, \quad s = 1,2,\ldots,r \tag{5.23}$$

where Z_k^*, W_s^* and Z_k^-, W_s^- are the ideal and anti-ideal solutions, respectively, of Equation 5.15.

The problem now is how to aggregate the objective functions in the fuzzy sense. Extend the concept of Bellman and Zadeh [1970], Zimmermann [1996] proposed two aggregate operators, the *min* operator, a non-compensatory operator, and the *product* operator, which is compensatory. However, the resulting problem by using the *product* operator is non-linear, which is generally difficult to solve. Thus, the *product* operator is seldom used. Using the *min* operator and follows similar argument as that used in obtaining Equation 5.14, the desired programming problem can be obtained. First notice that the final solution must satisfies all the membership functions of μ_k and μ_s in Equations 5.22 and 5.23. Thus, the conjunction and or the *min* operator must be used, we have:

$$\lambda = \min_i \mu_i(x) = \min_{k,s}(\mu_k, \mu_s) \tag{5.24}$$

where λ is the final membership function. The *min* operator can be replaced by inequality and, in addition, we would like to obtain the maximum value of the final membership function λ. Thus, the final programming problem is obtained from the above equations as:

$$\max \lambda \tag{5.25}$$

subject to:

$$\lambda \le \frac{Z_k(x) - Z_k^-}{Z_k^* - Z_k^-}, \quad k = 1,2,\ldots,\ell$$

$$\lambda \le \frac{W_s^- - W_s(x)}{W_s^- - W_s^*}, \quad s = 1,2,\ldots,r$$

$$Ax \le b$$

$$x \ge 0$$

The biggest disadvantage of (5.24) or (5.25), which is obtained by using the *min* operator, is due to the non-compensatory nature of the *min* operator. In other words, the results obtained by the *min* operator represent the worst situation and cannot be compensated by other members, which may be very good. To overcome this difficulty, a compensatory operator may be used. Lee and Li [1995] proposed the use of the *arithmetical average* operator and the programming problem becomes:

$$\max \ \xi = \frac{1}{\ell + r}\left(\sum_{k=1}^{\ell} \lambda_k + \sum_{s=1}^{r} \lambda_s\right) \tag{5.26}$$

subject to:

$$\lambda_k \le \frac{Z_k(x) - Z_k^-}{Z_k^* - Z_k^-}, \quad k = 1,2,\ldots,\ell$$

$$\lambda_s \le \frac{W_s^- - W_s(x)}{W_s^- - W_s^*}, \quad s = 1,2,\ldots,r$$

$$Ax \le b$$

$$x \ge 0$$

However, even with the compensatory operator, the results are frequently unbalanced. To obtain a balanced result, Lee and Li [1995] proposed the addition of a second phase by using the numerical results of the first phase. Assuming the solution obtained for the first phase is $(\lambda^{(1)}, x^{(1)})$, then the second phase is formulated as:

$$\max \ \xi = \frac{1}{\ell + r}\left(\sum_{k=1}^{\ell} \lambda_k + \sum_{s=1}^{r} \lambda_s\right). \tag{5.27}$$

subject to:

$$\lambda^{(1)} \le \lambda_k \le \frac{Z_k(x) - Z_k^-}{Z_k^* - Z_k^-}, \quad k = 1,2,\ldots,\ell$$

$$\lambda^{(1)} \le \lambda_s \le \frac{W_s^- - W_s(x)}{W_s^- - W_s^*}, \quad s = 1,2,\ldots,r$$

$$Ax \le b$$

$$x \geq 0$$

This two phase approach has been used by Lee and Li [1995] with satisfactory results.

5.4 Fuzzy Multiple-objective Programming with Fuzzy Parameters

The problems discussed in the previous section are crisp and not fuzzy ones. Thus, we are using the fuzzy concept to solve crisp problems, where the different objectives are competing. In this and the next sections, fuzzy problems with multiple-objectives will be considered. Many approaches have been proposed to solve this fuzzy problem. They can approximately be divided into two classes. The first class is based on the original crisp approach with the addition of fuzzy aspects and the second starts with the fuzzy concepts, which frequently give new alternative approaches. In this summary, two typical algorithms will be introduced in this and the next sections.

Depending on the degree of fuzziness, A fuzzy multiple-objective program can be classified into two categories:

1. fuzzy problem with both fuzzy objectives and fuzzy constraints but with crisp parameters;
2. fuzzy problem with fuzzy objectives, fuzzy constraints, and fuzzy parameters.

The first category can be solved by the approaches discussed in the previous section with minor modifications. This section will concentrate on the discussion of the second category, which can be expressed as:

$$\begin{aligned} &\max \tilde{Z}(x) = (\tilde{c}_1 x, \tilde{c}_2 x, \ldots, \tilde{c}_\ell x)^T \\ &\min \tilde{W}(x) = (\tilde{c}_1' x, \tilde{c}_2' x, \ldots, \tilde{c}_r')^T \end{aligned} \tag{5.28}$$

subject to:

$$x \in X = \{x \in R^n \mid \tilde{A}x * \tilde{b}, x \geq 0\}$$

where the parameters A, b and c with the ~ sign indicate that they are fuzzy parameters. For simplicity, we shall omit this ~ sign and it is understood that all the parameters in this section are assumed to be fuzzy. The fuzzy parameters c_k with $k = 1, 2, \ldots, \ell$, and c_s' with $s = 1, 2, \ldots, r$ are n-dimensional vectors, the fuzzy parameter b is an m-dimensional vector, and A is an $m \times n$ matrix.

Equation 5.28 has two types of fuzzy variables. The first type is the fuzzy goals resulting from the multi-objective compromise programming problem, which can be treated in a similar way as that discussed earlier. The second type is the fuzzy coefficients, which are fuzzy numbers instead of deterministic values.

In order to handle these fuzzy coefficients, let $(x)_\beta^\alpha$ be a solution of Equation 5.28. The symbol $\alpha \in [0, 1]$ denotes the level of possibility at which all fuzzy coefficients are feasible and $\beta \in [0, 1]$ denotes the grade of compromise to which the solution satisfies all of the fuzzy goals while the coefficients are at a feasible level α. By means of Bellman-Zadeh's rule of conjunction, the fuzzy parameters can be expressed as:

$$\alpha = \min_{k,s,i,j} (\mu_{ckj}, \mu_{csj}, \mu_{aij}, \mu_{bi}) \tag{5.29}$$

where k, s, i, and j are respectively ℓ-, r-, m-, and n-dimensional vectors. Equation 5.29 means that the feasibility of the system is equal to the possibility of the most impossible component in the system. Obviously, the optimal solution for a given value of α is reached when the following equation is true:

$$\mu_{ckj} = \mu_{csj} = \mu_{aij} = \mu_{bi} = \alpha \tag{5.30}$$

for all the subscripts.

Let $(P)_\alpha$ be the α-cut of a fuzzy number P, defined as:

$$(P)_\alpha = \{p \in S(P) | \mu_p(p) \geq \alpha\} \tag{5.31}$$

where $S(P)$ is the support of P. Let $(P)_\alpha^L$ and $(P)_\alpha^U$ be the lower and upper bounds respectively of the α-cut of P. Thus:

$$(P)_\alpha^L \leq (P)_\alpha \leq (P)_\alpha^U \tag{5.32}$$

For a given α, objectives Z_k to be maximized and W_s to be minimized can be replaced by the upper and lower bounds of their respective α-cuts. Thus, we have:

$$(Z_k)_\alpha^U = \sum_{j=1}^{n} (c_{kj})_\alpha^U x_j, \quad k = 1,2,\ldots,\ell \tag{5.33}$$

$$(W_s)_\alpha^L = \sum_{j=1}^{n} (c'_{sj})_\alpha^L x_j, \quad s = 1,2,\ldots,r \tag{5.34}$$

Now, consider the constraints in Equation 5.28. The general notation ★ in Equation 5.28 represents operators ≤, =, or ≥. The equality sign is equivalent to the pair of inequality signs, ≤ and ≥ [Lee and Li 1993]. Thus, only inequalities need to be considered. Using similar argument as that used for the objectives, we can express the constraints in terms of the upper and lower bounds. The constraints in Equation 5.28 can be expressed as:

$$\sum_{j=1}^{n} (a_{ij})_\alpha^L x_j \leq (b_i)_\alpha^U, \quad i = 1,2,\ldots,m_1, m_1+1, m_1+2,\ldots,m_1+m_2 \tag{5.35}$$

$$\sum_{j=1}^{n}(a_{ij})_{\alpha}^{U}x_j \geq (b_i)_{\alpha}^{L},\ i = m_1 + m_2 + 1, m_1 + m_2 + 2, \ldots, m_2 + m \tag{5.36}$$

where m_1 is the number of the "≤" operator, m_2 the "=" operator, and $(m - m_1 - m_2)$ the number of the "≥" operator.

Combining the results for the objectives and the constraints, Equation 5.28 can be represented by:

$$\max\ (Z_k)_{\alpha}^{U} = \sum_{j=1}^{n}(c_{kj})_{\alpha}^{U}x_j,\ k = 1,2,\ldots,\ell$$

$$\min\ (W_s)_{\alpha}^{L} = \sum_{j=1}^{n}(c_{sj})_{\alpha}^{L}x_j,\ s = 1,2,\ldots,r \tag{5.37}$$

subject to:

$$\sum_{j=1}^{n}(a_{ij})_{\alpha}^{L}x_j \leq (b_i)_{\alpha}^{U},\ i = 1,2,\ldots,m_1$$

$$\sum_{j=1}^{n}(a_{ij})_{\alpha}^{U}x_j \geq (b_i)_{\alpha}^{L},\ i = m_1, m_1 + 1 \ldots . m$$

$$x_j \geq 0,\ j = 1,2,\ldots,n$$

For a given value of α, the problem becomes a deterministic problem with multiple objectives. Thus, this resulting multi-objective problem can be reformulated in the same way as that discussed in Section 5.2.1. After this formulation, we finally have the following crisp programming problem:

$$\max\ \beta \tag{5.38}$$

subject to:

$$\beta \leq \mu_k^{\alpha}(Z_k) = \left[\sum_{j=1}^{n}(c_{kj})_{\alpha}^{U}x_j - (Z_k)_{\alpha}^{-}\right] \Big/ \left[(Z_k)_{\alpha}^{*} - (Z_k)_{\alpha}^{-}\right],\ k = 1,2,\ldots,\ell$$

$$\beta \geq \mu_s^{\alpha}(W_s) = \left[(W_s)_{\alpha}^{-} - \sum_{j=1}^{n}(c_{\alpha}')_{\alpha}^{L}x_j\right] \Big/ \left[(W_s)_{\alpha}^{-} - (W_s)_{\alpha}^{*}\right],\ s = 1,2,\ldots,r$$

$$\beta \in [0,1]$$

$$x \in X$$

where the inequality constraints in Equation 5.37 are represented by $x \in X$ and $(Z_k)_{\alpha}^{\star}$, $(W_s)_{\alpha}^{\star}$, and $(Z_k)_{\alpha}^{-}$, $(W_s)_{\alpha}^{-}$ are the ideal and anti-ideal solutions, respectively.

Since we must satisfy both α and β, by means of Bellman-Zadeh rule, the overall satisfaction for the solution $(x)_{\beta}^{\alpha}$ must be:

$$\lambda = \min \{\alpha, \beta) \tag{5.39}$$

Depending on the shape of the fuzzy number, Equation 5.38 can be further modified. The solution of the original problem Equation 5.28 can be obtained by solving Equations 5.38 and 5.39. In addition to the original unknown x variable, the problem represented by Equations 5.38 and 5.39 has two different types of additional unknown parameters, namely, the parameter α, which denotes the level of possibility for the fuzzy coefficients and the parameter β, which denotes the compromise between the different objectives. The problem represented by (5.38) and (5.39) is a non-linear problem and is not very easy to solve. Various solution schemes can be devised. Notice that if α is known, the problem can be solved easily. Thus, treating α as a parameter with different values, the problem can be solved repeatedly until the values of α and β become the same. This iterative scheme, which is like an exhaustive search procedure, can be carried out easily. The following example illustrates this procedure.

Example 5.3. Consider the following example, which was solved by Lee and Li [1995]:

$$\max \ \tilde{Z} = 10x_1 + \tilde{6}x_2$$
$$\min \ \tilde{W} = \tilde{1}x_1 + 1.5x_2$$

subject to:

$$\tilde{2}x_1 + 2x_2 \le \tilde{140}$$
$$x_1 \ge 0$$
$$x_2 \ge \tilde{8}$$

where all the fuzzy numbers are triangular fuzzy numbers, which are given as:

$$\tilde{6} = (4,6,8);\ \tilde{1} = (0,1,2);\ \tilde{2} = (1,2,3);\ \tilde{140} = (100,140,180);\ \tilde{8} = (3,8,10)$$

where the maximum possible value for the membership function is 1, which is achieved at the middle number.

Replace the fuzzy coefficients by their α-cuts, the above equation becomes:

$$\max \ (\tilde{Z})_\alpha = 10x_1 + (8 - 2\alpha)x_2$$
$$\min \ (\tilde{W})_\alpha = \alpha x_1 + 1.5x_2$$

subject to:

$$(1+\alpha)x_1 + 2x_2 \le 180 - 40\alpha$$

$$x_2 \geq 3 + 5\alpha$$
$$\alpha \in [0,1]$$
$$x_1 \geq 0$$
$$x_2 \geq 0$$

Using the above-mentioned exhaustive search procedure, this problem can be solved. The solution is:

$$\alpha = \beta = 0.67, x = (55.7, 6.35)$$
$$Z(x) = 599.8,\ W(x) = 46.7$$

5.5 Possibility Programming

Based on the possibility concept, which was discussed in Chapter 4, possibility programming can be formulated. The advantage of the approach is that it can directly use the knowledge of an expert and can translate linguistic expressions into a programming problem.

Luhandjula [1986] proposes a possibilistic linear programming, where the parameters or coefficients are characterized by possibility distributions. Buckley [1988, 1989] proposed an approach using triangular fuzzy numbers and discussed the difficulties in obtaining the possibility distribution due to non-linearity. Negi and Lee[1993] avoided the non-linearity problem by considering the actual situation during application, where it is frequently desirable to assign the desired or most appropriate possibility level according to the actual situation by the decision-maker. In fact, due to cost and other factors, we seldom require the system operat at the best possibility level. Once the possibility level is given *a priori*, the problem is reduced to a linear one. Negi and Lee [1993] used trapezoidal fuzzy number and strict exceedence possibility [Dubois and Prade 1988] to demonstrate their approach. In the present presentation, we shall follow the approach of Negi and Lee [1993].

One of the basic problem in fuzzy decision-making is the ranking or comparison of fuzzy numbers, which frequently have no linear order. Many different comparison approaches have been proposed [Zhu and Lee 1992, Chang and 1994]. Dubois and Prade [1988] proposed four indices based on the possibility concept. Only the strict exceedence possibility with trapezoidal fuzzy numbers will be summarized.

Consider two trapezoidal fuzzy numbers, B and R, which are represented by (b_1, b_2, b_3, b_4) and (r_1, r_2, r_3, r_4), respectively. The strict exceedence possibility is defined as:

$$Poss[\overline{B} \geq \overline{R}] = \begin{cases} 1, & if\ b_3 \geq r_4 \\ \delta, & if\ b_3 \leq r_4, b_4 \leq r_3 \\ 0, & if\ b_4 \leq r_3 \end{cases} \tag{5.40}$$

with

$$\delta = \frac{(b_4 - r_3)}{(b_4 - r_3) + (r_4 - r_3)} \tag{5.41}$$

5.5.1 Possibility Linear Programming

Consider the fuzzy linear programming:

$$\max \tilde{c}x \tag{5.42}$$

subject to:

$$\tilde{A}x \le \tilde{b}$$
$$x \le 0$$

Since the addition or subtraction of trapezoidal fuzzy numbers results in trapezoidal fuzzy numbers, the inequality constraints in the above equation can be treated as comparison of fuzzy numbers. Applying Equations 5.40 and 5.41 to these comparisons, we obtain the strict exceedence possibility for these inequalities:

$$Poss[x \in f_i] = \begin{cases} 1, & \text{if } b_{i3} \ge r_{i4} \\ \delta_i, & \text{if } b_{i3} \le r_{i4}, b_{i4} \ge r_{i3} \\ 0, & \text{if } b_{i4} \le r_{i3} \end{cases} \tag{5.43}$$

where

$$\delta_i = \frac{(b_{i4} - r_{i3})}{(b_{i4} - b_{i3}) + (r_{i4} - r_{i3})} \tag{5.44}$$

with

$$r_{i4} = \sum_{j=1}^{n} a_{ij4} x_j, \quad r_{i3} = \sum_{j=1}^{n} a_{ij3} x_j$$

where f_i represents the ith constraint in Equation 5.42. Since we are not interested in zero possibility, the decision space is restricted by the first and second equations of Equation 5.43. Thus, the only constraint that needs to be considered is:

$$b_{i4} \ge r_{i3}, \text{ for all } i \tag{5.45}$$

In order to satisfy all the constraints, we must take the minimum possibility among all the constraint possibilities, Thus:

$$\mathrm{Poss}[x \in f] = \min \{\mathrm{Poss}[x \in f_1], \mathrm{Poss}[x \in f_2], \ldots, \mathrm{Poss}[x \in f_m]\} = \min \{\delta_1, \delta_2, \ldots, \delta_m\} \tag{5.46}$$

The above is concerned with the constraints. We must also consider the objective function. The conditional possibility for the objective function z, given that the variables x's satisfy the strict exceedence possibility for the constraints, are:

$$Poss[Z = z|x] = \begin{cases} \theta_1, \ if \ \sum_{j=1}^{n} c_{j1}x_j \le z \le \sum_{j=1}^{n} c_{j2}x_j \\ 1, \ if \ \sum_{j=1}^{n} c_{j2}x_j \le z \le n\sum_{j=1}^{n} c_{j3}x_j \\ \theta_2, \ if \ \sum_{j=1}^{n} c_{j3}x_j \le z \le \sum_{j=1}^{n} c_{j4}x_j \end{cases} \tag{5.47}$$

where

$$\theta_1 = \frac{z - \sum_{j=1}^{n} c_{j1}x_j}{\sum_{j=1}^{n} c_{j2}x_j - \sum_{j=1}^{n} c_{j1}x_j}, \quad \theta_2 = \frac{\sum_{j=1}^{n} c_{j4}x_j - z}{\sum_{j=1}^{n} c_{j4}x_j - \sum_{j=1}^{n} c_{j3}x_j} \tag{5.48}$$

Since we want to maximize the objective function, the decision space we are interested is in the space of θ_2. The constraint for this space is the last constraint or the constraint for space θ_2:

$$\sum_{j=1}^{n} c_{j3}x_j \le z \le \sum_{j=1}^{n} c_{j4}x_j \tag{5.49}$$

Since the final unconditional possibility must satisfy both the objective function and all the constraints, The unconditional possibility must obey:

$$\mathrm{Poss}[Z{=}z] = \min \{\min (\delta_1, \delta_2, \ldots, \delta_m), \theta_2) = \min (\delta_1, \delta_2, \ldots, \delta_m, \theta_2) \tag{5.50}$$

subject to the constraints represented by Equation 5.45 and 5.49. From Equations 5.44 and 5.48, we can see that both the δ's and θ_2 are represented by non-linear equations. Thus, the minimization in Equation 5.50 is a non-linear problem.

This problem reduced to a linear one if the desired possibility are known and given. Assume that the desired possibility is given as α, then the problem reduced to:

$$\max z \tag{5.51}$$

subject to:

$$\delta_1, \delta_2, \ldots, \delta_m, \theta_2 \geq \alpha$$

$$b_{i4} \geq \sum_{j=1}^{n} a_{ij3} x_j, \quad i = 1,2,\ldots,m$$

$$\sum_{j=1}^{n} c_{j3} x_j \leq z \leq \sum_{j=1}^{n} c_{j4} x_j$$

$$x_j \geq 0, \quad j = 1,2,\ldots,n$$

Example 5.4. Consider the following fuzzy programming, which was solved by Negi and Lee[1993]:

$$\max \ z = (1, 2, 3, 4)\, x_1 + (2, 3, 4, 6)\, x_2$$

subject to:

$$(0, 1, 2, 3)\, x_1 + (1, 2, 3, 5)\, x_2 \leq (3, 5, 6, 8)$$
$$(1, 2, 3, 6)\, x_1 + (0, 1, 2, 3)\, x_2 \leq (2, 4, 6, 7)$$
$$x_1 \geq 0$$
$$x_2 \geq 0$$

For the first constraint, the strict exceedence possibility is:

$$Poss[x \in f_1] = \begin{cases} 1, & \text{if } 3x_1 + 5x_2 \leq 6 \\ \delta_1, & \text{if } 3x_1 + 5x_2 \geq 6,\ 2x_1 + 3x_2 \leq 8 \\ 0, & \text{if } 2x_1 + 3x_2 \geq 8 \end{cases}$$

where

$$\delta_1 = \frac{8 - 2x_1 - 3x_2}{2 + x_1 + 2x_2}$$

For the second constraint, the strict exceedence possibility is:

$$Poss[x \in f_2] = \begin{cases} 1, & \text{if } 6x_1 + 3x_2 \leq 6 \\ \delta_2, & \text{if } 6x_1 + 3x_2 \geq 6,\ 3x_1 + 2x_2 \leq 7 \\ 0, & \text{if } 3x_1 + 2x_2 \geq 7 \end{cases}$$

where

$$\delta_2 = \frac{7 - 3x_1 - 2x_2}{1 + 3x_1 + x_2}$$

With $\alpha = 0.5$ and by the use of strict exceedence possibility, the problem is reduced to the following crisp linear problem:

$$\max z$$

subject to:

$$
\begin{aligned}
&3x_1 + 4x_2 \le z \le 4x_1 + 6x_2 \\
&2x_1 + 3x_2 \le 8 \\
&3x_1 + 2x_2 \le 7 \\
&2.5x_1 + 4x_2 \le 7 \\
&4.5x_1 + 2.5x_2 \le 6.5 \\
&3.5x_1 + 5x_2 - z \ge 0 \\
&x_1, x_2 \ge 0
\end{aligned}
$$

The solution of this problem is: $z = 9.0213$, $(x_1, x_2) = (0.7234, 1.2979)$.

5.5.2 Multiple-objective Possibility Programming

A multiple-objective possibility linear problem can be expressed as:

$$\max \tilde{c}_k x, \quad k = 1,2,\ldots,p \tag{5.52}$$

subject to:

$$
\begin{aligned}
&\tilde{A}x \tilde{\le} \tilde{b} \\
&x \ge 0
\end{aligned}
$$

Where the "~" sign indicates that the variable is a fuzzy number.

The only difference between the current problem and that discussed in the previous section is that we now have p objective functions. Since all the objective functions require maximization, the decision space for the p objective functions must correspond to θ_2 in Equation 5.47. Thus, we have p equations correspond to Equation 5.49. If the decision-maker supplies a desired possibility level, Equation 5.52 can be reduced to:

$$\max z_k, \qquad k = 1, 2, \ldots, p \tag{5.53}$$

subject to:

$$
\begin{aligned}
&\delta_i \ge \alpha, && i = 1, 2, \ldots, m \\
&\theta_{k2} \ge \alpha, && k = 1, 2, \ldots, m \\
&b_{i4} \ge a_{ij3} x_j, && i = 1, 2, \ldots, m \\
&\sum_{j=1}^{n} c_{kj3} x_j \le z_k \le \sum_{j=1}^{n} c_{kj4} x_j, && k = 1, 2, \ldots, p \\
&x_j \ge 0, && j = 1, 2, \ldots, n
\end{aligned}
$$

Equation 5.53 is a linear multiple-objective problem with $(n + p)$ unknowns and with $(2m + 3p)$ constraints. This problem can be solved easily.

Instead of strict exceedence possibility, the three other possibilities defined by Dubois and Prade [1988], or other definition of possibility, can also be used. Although the final equations are different, the procedure to obtain the equations is essentially the same. The following example uses the exceedence possibility as defined by Dubois and Prade [1988].

Example 5.5. Consider the following possibility problem, whose parameters are trapezoidal fuzzy numbers:

$$\max f_1 = (1.5, 2, 2, 2.5)\, x_1 + (-1.5, -1, -1, -0.5)\, x_2$$
$$\max f_2 = (0.8, 1, 1, 1.4)\, x_1 + (1.5, 2, 2.5, 2.5)\, x_2$$

subject to:

$$(2.5, 3, 3, 3.5)\, x_1 + (-5.5, -5, -5, -4.5)\, x_2 \leq (10, 12, 15, 26)$$
$$(2.8, 3, 3.2, 3.5)\, x_1 + (-1.5, -1, -1, -0.5)\, x_2 \leq (15, 20, 25, 35)$$
$$(2.75, 3, 3.2, 3.2)\, x_1 + (0.8, 1, 1, 1.3)\, x_2 \leq (20, 25, 25, 42)$$
$$(2.5, 3, 3, 3.5)\, x_1 + (3.6, 4, 4, 4.4)\, x_2 \leq (30, 32, 35, 50)$$
$$(0.8, 1, 1, 1.2)\, x_1 + (2.6, 3, 3, 3.4)\, x_2 \leq (14, 15, 16, 25)$$
$$x_1, x_2 \geq 0$$

Using exceedence possibility, the above problem reduced to:

$$\min\{\theta_{12}, \theta_{22}, \delta_1, \delta_2, \ldots, \delta_5\}$$

subject to:

$$\theta_{12} = (2.5\, x_1 - 0.5\, x_2 - f_1) / (0.5\, x_1 + 0.5\, x_2) \geq \alpha$$
$$\theta_{22} = (1.4\, x_1 + 2.5\, x_2 - f_2) / (0.4 - x_1) \geq \alpha$$
$$2\, x_1 - x_2 \leq f_1 \leq 2.5\, x_1 - 0.5\, x_2$$
$$x_1 + 2.5\, x_2 \leq f_2 \leq 1.4\, x_1 + 2.5\, x_2$$
$$\delta_1 = (26 - 2.5\, x_1 + 5.5\, x_2) / (11 + 0.5\, x_1 + 0.5\, x_2) \geq \alpha$$
$$15 \leq 3\, x_1 - 5\, x_2$$
$$26 \geq 2.5\, x_1 - 5.5\, x_2$$
$$\delta_2 = (35 - 2.8\, x_1 + 1.5\, x_2) / (10 + 0.2\, x_1 + 0.5\, x_2) \geq \alpha$$
$$25 \leq 3\, x_1 - x_2$$
$$35 \geq 2.8\, x_1 - 1.5\, x_2$$
$$\delta_3 = (42 - 2.75\, x_1 - 0.8\, x_2) / (17 + 0.25\, x_1 + 0.2\, x_2 \geq \alpha$$
$$25 \leq 3\, x_1 + x_2$$
$$42 \geq 2.75\, x_1 + 0.8\, x_2$$
$$\delta_4 = (50 - 2.5\, x_1 - 3.6\, x_2) / (15 + 0.5\, x_1 + 0.4\, x_2) \geq \alpha$$
$$35 \leq 3\, x_1 + 4\, x_2$$
$$50 \geq 2.5\, x_1 + 3.6\, x_2$$
$$\delta_5 = (25 - 0.8x_1 - 2.6x_2) / (9 + 0.2x_1 + 0.4x_2) \geq \alpha$$
$$16 \leq x_1 + 3x_2,$$
$$25 \geq 0.8x_1 + 2.6x_2$$

$$\delta_1, \delta_2, \delta_3, \delta_4, \delta_5 \in [0, 1]$$
$$\theta_{12}, \theta_{22} \in [0, 1]$$
$$\alpha \in [0, 1]$$
$$x_1, x_2 \geq 0$$

This problem was solved by a trial-and-error approach. The solution obtained is $x = (9.1, 2.4)$ with $f = (15.9, 17.74)$. The degree of satisfaction is $\alpha = 0.767$.

Chapter 6

Fuzzy Interactive Multi-level Decision Making

The duo-ploy or multi-ploy decision-making process discussed in Chapters 2 and 3 is too rigid and is not, in general, suited for an organization such as a manufacturing or service company. The decision-making process in a well-organized company requires better control of the top level and certainly cannot allow the ambiguity which happens if the rational reaction sets of the lower levels are not singletons. Furthermore, the lower decision should not be the final decision and certainly should allow the inspection of the higher levels. In fact, there should have mutual consultations so that the best objective for the overall company can be achieved. It is true that, depending on the management style of the top level and the organization of the company, different degrees of control of the higher levels to the lower levels may be carried out. Because of these different degrees of control, different or flexible interactions between the different levels should be allowed. Although some of the algorithms discussed in Chapters 2 and 3 can be extended or modified to allow this flexibility, the philosophy of the decision process would be completely different and, furthermore, the computational requirements would increase tremendously. Since the algorithms are already limited to reasonably medium sized problems, the algorithms resulting from this direct modification or extension would not be efficient.

In this chapter, fuzzy interactive decision making, which allows various different degrees of control and thus is ideally suited for a manufacturing or service company, will be investigated. In addition, as we shall see later, that this approach has the following advantages:

1. The problem is simplified and, at the same time, the representation is more realistic. This is because we are treating a fuzzy and not well defined problem as it is, or, not try to represent it more accurately than it really is;
2. The use of fuzzy membership functions to represent the desires or goals of the DMs in different levels offers exceptional flexibility for the interactive decisions proposed;

3. Since the problem is much more simplified, it is much easier to solve by using fuzzy multiple-objective decision-making algorithms.

Another example is the solution process of the kth-best algorithm. It is certainly not reasonable to assume the solution is at the corner point. As mentioned before, this unreasonable requirement resulted in two different optimal solutions for the same problem in Examples 2.1 and 2.3. These different results are due to the fact that the optimum depends on which DM moves first in the search procedure. Furthermore, according to the concept of multi-criteria decision making, the Pareto optimum does not necessarily lie at the vertex. All the best results require are that a compromise non-dominated solution is obtained based on the DM's preferences. From this standpoint, the last few steps of the search procedure, where the winner of the two competing neighboring vertices is decided, are not needed extra computations. We are trying to force a decision instead of the more reasonable compromise decision by consultation.

Instead of searching through the vertices as in the kth-best or other similar algorithms, a supervised search procedure, which, hopefully, will generate a non-dominated satisfactory solution for the multi-level decision-making process, is proposed in this chapter. The use of fuzzy membership functions, which simplify the representation and the computations considerably, to represent the compromises between the different levels implicitly forms the basic idea of the approach.

From another standpoint, fuzzy interactive decision making is essentially a continuation of the duo- or multi-ploy search procedure. However, in order to carry out this continuation, considerable simplification in both the representation of the process and in the computational requirements are needed. The proposed interactive decision approach provides these simplifications by exploring the basic vagueness and the well-defined nature of most practically encountered hierarchy organizations. In other words, instead of taking the lower decision as the final decision in the multi-ploy approach, the search process continues until all levels involved are satisfied. However, due to the complexity of the algorithms, this continuation cannot be carried out easily with the multi-ploy type algorithms.

In this chapter, the simplest bi-level process will be discussed first. Other structurally more complicated processes such as the bi-level with multi-followers and multi-level processes will be presented afterward. To emphasize the computational aspects, numerical examples will be used to illustrate the approaches.

6.1 Fuzzy Bi-level Interactive Decision Making

Instead of the duo-ploy type programming approach, consider the fuzzy interactive decision-making process applied to a bi-level organization with a single follower. The upper-level DM specifies the preferred values of his or her control variables and goals with certain amount of tolerance. This information is represented implicitly by the use of membership functions and passed to the lower-level DM. The lower-level DM obtains his or her optimum based on goals and preferences of the upper level and then presents the results to the upper level. If the upper level agrees with the proposed solutions, a final decision is reached and, for the convenience of description, this decision or solution will be referred to as a satisfactory solution. If he or she rejects this proposal, the DMs in both levels will need to re-evaluate and changes the goals and decisions as well as their corresponding tolerances. This mutually interactive process is continued until a satisfactory solution is reached. This strategy is very flexible. Since the DMs in both levels first seek their optimal solutions in isolation, it does not violate the non-cooperate idea. However, the strategy does require a certain degree of coordination between the different levels.

Mathematically, the upper-level DM first solves the following problem:

$$\max f_1(x_1, x_2) = c_{11}^{\mathrm{T}} x_1 + c_{12}^{\mathrm{T}} x_2 \tag{6.1}$$

subject to:

$$(x_1, x_2) \in F_2 = \{(x_1, x_2) \mid A_1 x_1 + A_2 x_2 \le b,\ x_1,\ x_2 \ge 0\}$$

the solution of which is assumed as (x_1^U, x_2^U, f_1^U). Independently, the lower-level DM solves:

$$\max f_2(x_1, x_2) = c_{21}^{T} x_1 + c_{22}^{T} x_2 \tag{6.2}$$

subject to:

$$(x_1, x_2) \in F_2 = \{(x_1, x_2) \mid A_1 x_1 + A_2 x_2 \le b, x_1, x_2 \ge 0\}$$

the solution of which is assumed as (x_1^L, x_2^L, f_2^L). If $(x_1^U, x_2^U) = (x_1^L, x_2^L)$, an optimal or preferred solution is reached. However, the two solutions are usually different because of the conflicting nature of the two objectives. The upper-level DM understands that using the optimal decision x_1^U as a control factor for the lower-level DM is obviously not practical. It is more reasonable to have some tolerance that gives the lower-level DM a wider feasible region to search for his or her optimum. The range of the decision for the vector x_1 can be "around x_1^U" with its maximum tolerance p_1. In other words, the most preferred decision is x_1^U and the worst acceptable decision is at $(x_1^U - p_1)$ or $(x_1^U + p_1)$. The satisfaction or preference can be assumed as linearly increasing within the interval of $[x_1^U - p_1, x_1^U]$ and linearly decreasing within $[x_1^U, x_1^U + p_1]$. Decisions outside the interval $[x_1^U - p_1, x_1^U + p_1]$ are not acceptable. This information can be formulated as the following membership function of the fuzzy set theory:

$$\mu_{x1}(x_1) = \begin{cases} [x_1 - (x_1^U - p_1)] / p_1, & \text{if } x_1^U - p_1 \le x_1 \le x_1^U \\ [(x_1^U + p_1) - x_1] / p_1, & \text{if } x_1^U \le x_1 \le x_1^U + p_1 \\ 0, & \text{otherwise} \end{cases} \tag{6.3}$$

The upper-level DM must also reset his or her goal with some tolerance. The upper level may adopt a leeway of f_1'. Any value of the goal $f_1 > f_1^U$ is absolutely acceptable and all values of $f_1 < f_1'$ are absolutely unacceptable. The satisfaction or preference within the interval $[f_1', f_1^U]$ can be assumed as linearly increasing. The membership function can now be assumed as follows:

$$\mu_{f1}(f_1(x)) = \begin{cases} 1, \text{ if } f_1(x) \geq f_1^U \\ [f_1(x) - f_1'] / [f_1^U - f_1'], \text{ if } f_1' \leq f_1(x) \leq f_1^U \\ 0, \text{ if } f_1(x) \leq f_1' \end{cases} \tag{6.4}$$

As an initial approximation, the value of f_1' is usually set equal to $f_1(x_1^L, x_2^L)$.

The lower-level DM now optimizes her objective function under the new constraints of "x_1 is about x_1^U" and "f_1 is somehow near to or greater than f_1^U", which are modeled by the membership functions (Equations 6.3 and 6.4). Using Equations 6.2-6.4, the problem of the lower level becomes:

$$\max_{x2} f_2(x_1, x_2) = c_{21}^T x_1 + c_{22}^T x_2 \tag{6.5}$$

subject to:

$$\begin{aligned} &A_1 x_1 + A_2 x_2 \leq b \\ &f_1(x) \geq f_1' \\ &x_1 \cong x_1^U \\ &x_1, x_2 \geq 0 \end{aligned}$$

or

$$\max_{x2} f_2(x_1, x_2) = c_{21}^T x_1 + c_{22}^T x_2 \tag{6.6}$$

subject to:

$$\begin{aligned} &A_1 x_1 + A_2 x_2 \leq b \\ &\mu_{f1}(f_1(x)) \geq \alpha \\ &\mu_{x1}(x_1) \geq \beta \\ &\alpha \in [0, 1] \\ &\beta \in [0, 1] \\ &x_1, x_2 \geq 0 \end{aligned}$$

where scalars α and β are the minimum acceptable degrees of satisfaction for the objective $f_1(x)$ and the decision x_1, respectively. These feasible ranges are constrained by $\mu_{f1}(f_1(x))$ and $\mu_{x1}(x_1)$. The lower-level DM can analyze the various solutions corresponding to the upper-level DM's satisfactory levels α and β. These acceptable degrees of satisfaction are generally decided by the DMs.

In addition to the tolerance of the DM of the upper level, the lower-level DM may also have tolerance or preference level. This preference can again be represented by membership function so that he or she can rate the satisfaction of each potential solution. Let this membership function be defined as:

$$\mu_{f2}(f_2(x)) = \begin{cases} 1, \text{ if } f_2(x) \geq f_2^L \\ [f_2(x) - f_2'] / [f_2^L - f_2'], \text{ if } f_2' \leq f_2(x) \leq f_2^L \\ 0, \text{ if } f_2(x) \leq f_2' \end{cases} \tag{6.7}$$

where $f_2' = f_2(x_1^T)$. Following the discussions made in Chapter 5, the lower-level DM must maximize his or her objective function within the interval $[f_2', f_2^L]$. Thus, we have:

$$\max \delta = \mu_{f2}(f_2(x)) \tag{6.8}$$

subject to:

$$\begin{aligned} &A_1 x_1 + A_2 x_2 \leq b \\ &\mu_{f1}(f_1(x)) \geq \alpha \\ &\mu_{x1}(x) \geq \beta \\ &\alpha \in [0, 1] \\ &\beta \in [0, 1] \\ &x_1, x_2 \geq 0 \end{aligned}$$

or

$$\max \delta \tag{6.9}$$

subject to:

$$\begin{aligned} &A_1 x_1 + A_2 x_2 \leq b \\ &\mu_{f1}(f_1(x)) \geq \alpha \\ &\mu_{x1}(x_1) \geq \beta \\ &\mu_{f2}(f_2(x)) \geq \delta \\ &\alpha \in [0,1] \\ &\beta \in [0, 1] \\ &\delta \in [0, 1] \\ &x_1, x_2 \geq 0 \end{aligned}$$

where δ is the degree of satisfactory of the lower-level DM's objective function. In order to satisfy both upper and the lower levels or all the degrees of satisfaction, the solution must be the minimum of $\{\alpha, \beta, \delta\}$, or min $\{\alpha, \beta, \delta\}$. Let:

$$\lambda = \min \{\alpha, \beta, \delta\} \tag{6.10}$$

where λ is a fuzzy number resulting from the intersection of the three membership functions. In order to obtain the maximum degree of satisfaction of this fuzzy number, we must have:

$$\max \lambda = \max (\min \{\alpha, \beta, \delta\}) \tag{6.11}$$

subject to:

$$\begin{aligned}
&A_1 x_1 + A_2 x_2 \leq b \\
&\mu_{f1}(f_1(x)) \geq \alpha \\
&\mu_{x1}(x_1) \geq \beta \\
&\mu_{f2}(f_2(x)) \geq \delta \\
&\alpha \in [0, 1] \\
&\beta \in [0, 1] \\
&\delta \in [0, 1] \\
&x_1, x_2 \geq 0
\end{aligned}$$

Equation 6.10 can also be written as:

$$\begin{aligned}
&\mu_{f1}(f_1(x)) \geq \alpha \geq \lambda \\
&\mu_{x1}(x_1) \geq \beta \geq \lambda \\
&\mu_{f2}(f_2(x)) \geq \delta \geq \lambda
\end{aligned}$$

Using the above three inequalities, the three degrees of satisfaction, α, β and δ, in Equation 6.11 can be eliminated. Thus, Equation 6.11 becomes:

$$\max \lambda \tag{6.12}$$

subject to:

$$\begin{aligned}
&A_1 x_1 + A_2 x_2 \leq b \\
&\mu_{f1}(f_1(x)) \geq \lambda \\
&\mu_{x1}(x_1) \geq \lambda \\
&\mu_{f2}(f_2(x)) \geq \lambda \\
&\lambda \in [0, 1] \\
&x_1, x_2 \geq 0
\end{aligned}$$

or

$$\max \lambda \tag{6.13}$$

subject to:

$$\begin{aligned}
&A_1 x_1 + A_2 x_2 \leq b \\
&[f_1(x) - f_1'] / [f_1^U - f_1'] \geq \lambda \\
&[(x_1^U + p_1) - x_1] / p_1 \geq \lambda \\
&[x_1 - (x_1^U - p_1)] / p_1 \geq \lambda \\
&[f_2(x) - f_2'] / [f_2^L - f_2'] \geq \lambda \\
&\lambda \in [0, 1] \\
&x_1, x_2 \geq 0
\end{aligned}$$

where the constraint on x_1 is a vector, the dimension of which depends on the number of decisions of the upper level. Equation 6.11 is a max-min programming problem proposed by Zimmermann [1978] based on the suggestions of Bellman and Zadeh [1970].

If the upper-level DM is satisfied with the solution of Equation 6.11, a satisfactory solution is obtained. If not satisfied, the upper level should provide new membership functions for the control variable and the objective to the lower level. This process is continued until a satisfactory solution is reached. Combined with the set of control decisions and goals with tolerances, this solution becomes a satisfactory solution for Equations 6.1 and 6.2. Figure 6.1 shows the flow chart of this proposed procedure, which can be summarized in the following three steps:

Step 1. The upper and lower level's DMs solve their problems independently by solving Equations 3.13 and 3.14. If these two solutions coincide, the optimal or preferred solution of the system is obtained. Stop. Otherwise, go to Step 2.

Step 2. The upper-level DM decides her tolerances on the goal and the decisions in terms of membership functions. Meanwhile, the lower-level DM also decides his tolerance on the goal in terms of membership function. These membership functions will serve as extra constraints in forming the auxiliary problem, Equation 6.13.

Step 3. Solve the auxiliary problem. If the DMs at each level are satisfied with the solution, a compromise solution is reached. Stop. Otherwise, got to Step 2 to obtain newly adjusted membership functions.

To illustrate the approach, let us consider the following example.

Example 6.1. Problem of balancing export and import solved in Examples 2.1 and 2.3 is again solved by the present approach. The problem is reproduced in the following:

$$\max_{x1} f_1 = 2x_1 - x_2 \qquad \text{(upper level) (6.14a)}$$

where x_2 solves:

$$\max_{x2} f_2 = x_1 + 2x_2 \qquad \text{(lower level) (6.14b)}$$

subject to:

$$\begin{aligned} 3x_1 - 5x_2 &\le 15 \\ 3x_1 - x_2 &\le 21 \\ 3x_1 + x_2 &\le 27 \\ 3x_1 + 4x_2 &\le 45 \\ x_1 + 3x_2 &\le 30 \\ x_1, x_2 &\ge 0 \end{aligned}$$

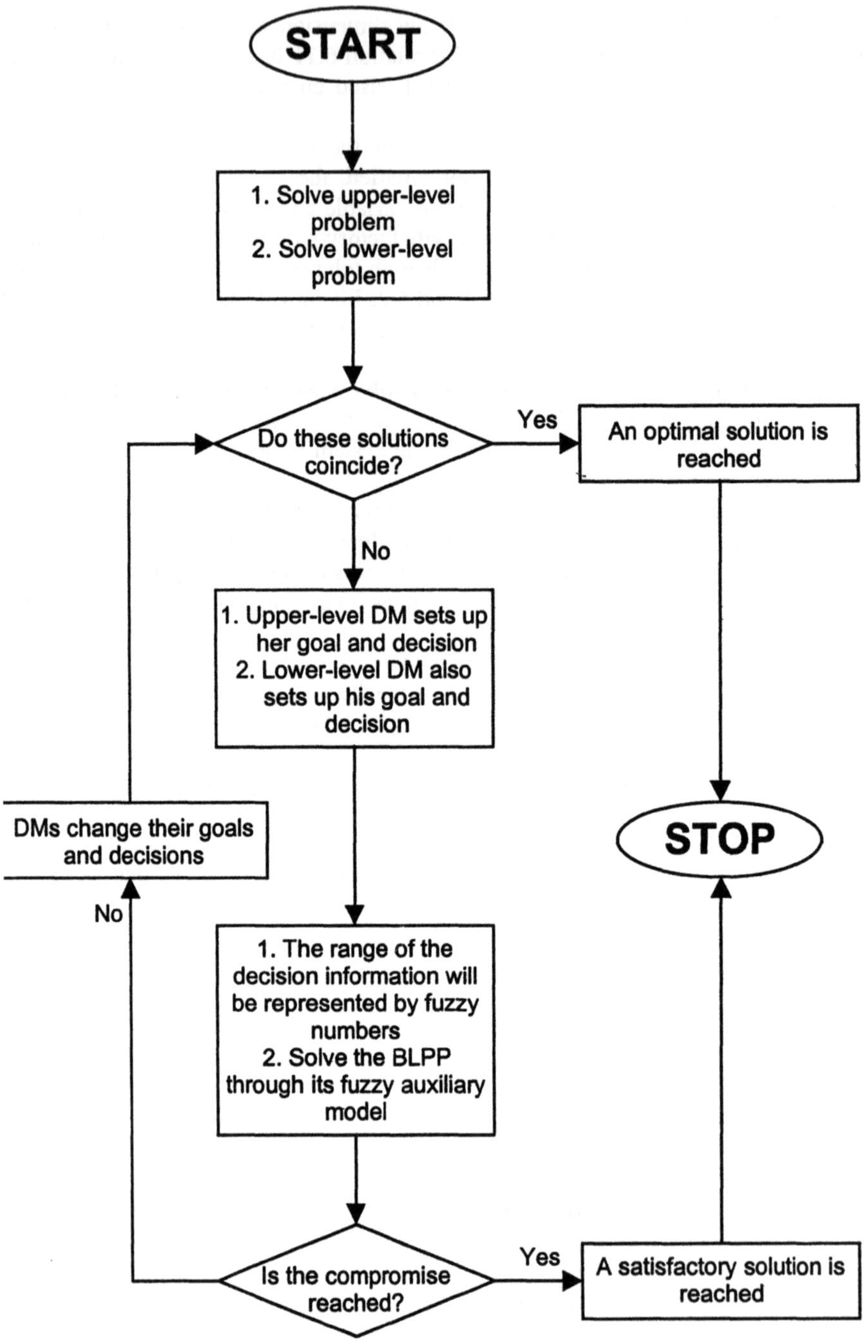

Figure 6.1. Fuzzy interactive decision making

The proposed approach first solves Equations 6.14a and 6.14b separately subject to the above constraint. The solutions obtained are $(x_1^U, x_2^U) = (7.5, 1.5)$ with $f_1^U = 13.5$ and $(x_1^L, x_2^L) = (3, 9)$ with $f_2^L = 21$. Using this solution as reference, set $f_1^U = 13.5$ and assume $f_1' = 0$ and $f_2' = 10.5$. Let the upper-level DM's decision variable x_1 be around 7.5 with the negative- and positive-side tolerances equal to 4.5 and 0.5, respectively. Using Equations 6.3, 6.4 and 6.7, the membership functions $\mu_{f1}(\bullet)$, $\mu_{x1}(\bullet)$ and $\mu_{f2}(\bullet)$ are obtained. The lower-level DM then solves the following problem:

$$\max \lambda = \max (\min \{\alpha, \beta, \delta\})$$

subject to:

$$\begin{aligned}
&x \in X \\
&\mu_{f1}(f_1(x)) = (f_1 - 0) / (13.5 - 0) \geq \alpha \\
&\mu_{x1}(x_1) = (x_1 - 4.5) / (7.5 - 4.5) \geq \beta \\
&\mu_{x1}(x_1) = (8 - x_1) / (8 - 7.5) \geq \beta \\
&\mu_{f2}(f_2(x)) = (f_2 - 10.5) / (21 - 10.5) \geq \delta \\
&x_1, x_2 \geq 0 \\
&\alpha, \beta, \delta \in [0, 1]
\end{aligned}$$

Using the same argument as that used in obtaining Equation 6.12, we can obtain the following equation:

$$\max \lambda$$

subject to:

$$\begin{aligned}
&x \in X \\
&x_1 \geq 4.5 + 3\lambda \\
&x_1 \leq 8 - 0.5\lambda \\
&f_1 = 2x_1 - x_2 \geq 13.5\lambda \\
&f_2 = x_1 + 2x_2 \geq 10.5\lambda + 10.5 \\
&\lambda \in [0, 1] \\
&x_1, x_2 \geq 0
\end{aligned}$$

the compromise solution of which is $f^* = (f_1^*, f_2^*) = (9.29, 17.72)$ with $x^* = (x_1^*, x_2^*) = (7.26, 5.23)$. This solution is a point between vertices D and C in Figures 2.1 and 2.2. The overall satisfaction of the present solution is $\lambda = 0.688$. The satisfaction levels are $(\alpha, \beta, \delta) = (0.688, 0.947, 0.688)$. If the upper-level DM's total satisfactory level is $\lambda_1 = \min \{\alpha, \beta\}$, then the satisfaction levels for upper and lower levels are $\lambda_1 = 0.688$ and $\lambda_2 = \delta = 0.688$. On the other hand, the kth-best solution gave f = (13, 14) with x = (8, 3), which has a satisfaction level of $(\alpha, \beta, \delta) = (0.963, 0, 0.333)$ and thus $\lambda_1 = 0$ and $\lambda_2 = 0.333$. Thus, the present solution is better balanced than that of the kth-best approach.

If the solution is not satisfactory to any one of the DMs, adjustment of the tolerances can be made. Again, this process can be continued until a solution, which is satisfactory to all the DMs is reached.

6.2 Fuzzy Bi-level Interactive Decision Making with Multi-followers

The duo-ploy type decision-making process with multi-followers, or bi-level distributed programming problem (BLDPP), was discussed in Chapter 3. However, for the decision process in most organizations, more cooperation than the duo-ploy type is needed. The fuzzy interactive decision-making process discussed in the last section can be easily extended to the bi-level organization with multi-followers. For the easy of discussion, the bi-level decentralized problem with multi-followers, Equation 3.1, is reproduced in the following:

$$\max_{x11} f_{11}(x) = \textstyle\sum_j c_{11j}^T x_j \qquad \text{(upper level) (6.15a)}$$

where $x_{21}, x_{22}, ..., x_{2p}$ solve:

$$\begin{cases} \max_{x21} f_{21}(\mathrm{x}) = \sum_j c_{21j}^T x_j \\ \ldots \\ \max_{x2p} f_{2p}(x) = \sum_j c_{2pj}^T x_j \end{cases} \qquad \text{(lower level) (6.15b)}$$

subject to:

$$\textstyle\sum_{k,i} A_{ki}\, x_{ki} \le b,\ k = 1, 2 \text{ and } i = 1, 2, ..., s_k$$
$$x_j \ge 0, j = 1, 2, ..., n$$

For convenience, let X represent the above constraint set.

First solve each DM's problem separately subject to the constraints of Equations 6.15b:

$$f_{ki}^* = f_{ki}(x_{ki}^*) = \max_{x\in X} f_{ki}(x) \qquad (6.16)$$

for all k and i, where the superscript * indicates the actual solution, or the individual optimum values obtained.

Then, follow the same argument used before, that is, the minimum satisfaction level would satisfy all the DMs. Or, let:

$$f_{ki}' = \min_{k,i} f_{ki}(x_{ki}^*) \qquad (6.17)$$

where the minimization is with respect to all the degrees of satisfaction of all the DMs. If the tolerance vector, $\boldsymbol{p}_1$, for the upper DM is given, then the membership functions, $\mu_{x1}(x_1)$ and $\mu_{ki}(f_{ki}(x))$, can be formulated in the same way as before. Thus the lower-level DMs' problem becomes:

$$\begin{cases} \max_{x21} f_{21}(x) = \sum_j c_{21j}^T x_j \\ \ldots \\ \max_{x2p} f_{2p}(x) = \sum_j c_{2pj}^T x_j \end{cases} \tag{6.18}$$

subject to:

$$\begin{aligned} & x \in X \\ & \mu_{f1}(f_1(x)) \geq \alpha \\ & \mu_{x1}(x_1) \geq \beta \\ & \alpha \in [0, 1] \\ & \beta \in [0, 1] \end{aligned}$$

Introducing the lower-level tolerance function and eliminate the degrees of satisfaction vector, we have:

$$\max \lambda \tag{6.19}$$

subject to:

$$\begin{aligned} & x \in X \\ & \mu_{x1}(x_1) \geq \lambda \\ & \mu_{f1}(f_1(x)) \geq \lambda \\ & \mu_{f2i}(\mathrm{f}_{2i}(x)) \geq \lambda_i \\ & \lambda \in [0, 1] \\ & \lambda_i \in [0, 1] \end{aligned}$$

with $i = 1, 2, \ldots, p$. Substituting the definition of the membership functions into the above equation, we finally have:

$$\max \lambda \tag{6.20}$$

subject to:

$$\begin{aligned} & x \in X \\ & [(x_1^U + p_1) - x_1] / p_1 \geq \lambda \\ & [x_1 - (x_1^u - p_1)] / p_1 \geq \lambda \\ & \mu_{f1}(f_1(x)) = [f_1(x) - f_1'] / [f_1^T - f_1'] \geq \lambda \\ & \mu_{f2i}(f_{2i}(x)) = [f_{2i}(x) - f_{2i}'] / [f_{2i}^T - f_{2i}'] \geq \lambda_i \\ & \lambda \in [0, 1] \\ & \lambda_i \in [0, 1],\ i = 1, 2, \ldots, p \end{aligned}$$

Solving Equation 6.20, the degrees of satisfaction of the various DMs can be obtained. If these degrees of satisfactions do not satisfy all the DMs, the membership functions or the tolerances of the DMs are re-adjusted. Again, this process can be continued until a satisfactory solution is reached.

In order to obtain the initial solution of Equation 6.16, notice that the problem such as Equation 6.15 really represents two problems, the top-level problem and the bottom-level problem. By itself, the top-level problem is fairly simple. However, since it is imbedded in the lower level, it would be ideal if we could solve the lower level first. Since all the divisions in the bottom level have the same position regarding to the decision sequence, or, they make the decisions simultaneously and in isolation based on the top level's decision, the problem of the bottom level with a given $\boldsymbol{x}_1$ is no longer a hierarchy. In fact, the bottom level, by itself, is a one-level optimization problem. If we allow mutual interactions, the bottom-level decision process is essentially a multi-objective optimization problem. Thus, by using the fuzzy multi-objective decision-making algorithms discussed in Chapter 5, the bottom-level problem of Equation 6.16 can be solved.

The approach is illustrated by solving the following example.

Example 6.2. Consider the following bi-level problem with three followers, which was solved by Anandalingdam [1988] and also solved in Example 3.1

$$\max_{x1} f_1 = x_1 + y_1 + 2y_2 + y_3 \tag{6.21}$$

where y_1, y_2, and y_3 solve:

$$\begin{cases} \max_{y1} f_{21} = -x_1 + 3y_1 - y_2 - y_3 \\ \max_{y2} f_{22} = -x_1 - y_1 + 3y_2 - y_3 \\ \max_{y3} f_{23} = -x_1 - y_1 - y_2 + 3y_3 \end{cases}$$

subject to:

$$\begin{aligned} &3x_1 + 3\,y_1 \le 30 \\ &2x_1 + y_1 \le 20 \\ &y_2 \le 10 \\ &y_2 + y_3 \le 15 \\ &y_3 \le 10 \\ &x_1 + 2y_1 + 2y_2 + y_3 \le 40 \\ &x_1, y_1, y_2, y_3 \ge 0 \end{aligned}$$

Table 6.1. Solutions of individual objectives, Example 6.2

Vertex	f1	f21	f22	f23
A (10, 0, 10, 5)	35*	-25	15	-5
B (5, 5, 10, 5)	35*	-5	15	-5
C (0, 10, 0, 0)	10	30*	-10	-10
D (0, 0, 10, 0)	20	-10	30*	-10
E (0, 0, 0, 10)	10	-10	-10	30*
Fuzzy range	(10, 35)	(0, 30)	(0, 30)	(0, 30)

Individual optimal solutions of the top and bottom-level DMs subject to the constraints of Equation 6.21 are listed in Table 6.1, where the "*" represents the optimum solution. Using these solutions as basis, we can assume that $f_{11}' = 0$ and

$f_{21}' = f_{22}' = f_{23}' = 0$. Values within the interval [6, 10] for x_1 are fully satisfied to the upper-level DM. The negative and positive side tolerances for x_1 are 5 and 0, respectively. The auxiliary equation now becomes:

$$\max \lambda$$

subject to:

$$\begin{aligned}
&\boldsymbol{x} \in \boldsymbol{X} \\
&(x_1 + y_1 + 2y_2 + y_3 - 10) / (35 - 10) \geq \alpha \\
&(x_1 - 0) / (5 - 0) \geq \beta \\
&x_1 \leq 10 \\
&(-x_1 + 3y_1 - y_2 - y_3 - 0) / (30 - 0) \geq \gamma_1 \\
&(-x_1 - y_1 + 3y_2 - y_3 - 0) / (30 - 0) \geq \gamma_2 \\
&(-x_1 - y_1 - y_2 + 3y_3 - 0) / (30 - 0) \geq \gamma_3
\end{aligned}$$

Solving the above equation, we obtain the satisfactory solution $\boldsymbol{f} = (f_1, f_{21}, f_{22}, f_{23}) = (31.07, 6.43, 6.43, 6.43)$ with $\boldsymbol{x} = (x_1, y_1, y_2, y_3) = (1.07, 7.5, 7.5, 7.5)$. The satisfaction levels are (0.21, 0.21, 0.21, 0.21) for all the DMs. The solution of Example 3.1 and also Anandalingam's solution [1988] is $\boldsymbol{f} = (35, -5, 15, -5)$ with $\boldsymbol{x} = (5, 5, 10, 5)$, whose satisfaction is (1, 0, 0.5, 0). Thus, the proposed solution is much more reasonable and more evenly distributed.

6.3 Fuzzy Multi-level Interactive Decision Making

A multi-level decision-making process in an organization, in which there are K nested hierarchy levels, can be represented as:

$$\max_{x1} f_1(x) = \textstyle\sum_j c_{1j}^T x_j \qquad (1^{st} \text{ level}) \quad (6.22)$$

where $x_2, x_3, \ldots, x_K$ solve:

$$\max_{x2} f_2(x) = \textstyle\sum_j c_{2j}^T x_j \qquad (2^{nd} \text{ level})$$

where $x_3, \ldots, x_K$ solve:

...

where x_K solves:

$$\max_{xk} f_K(x) = \textstyle\sum_j c_{Kj}^T x_j \qquad (k^{th} \text{ level})$$

subject to:

$$\sum_k A_k x_k \le b,\ k = 1, 2, ..., K$$
$$x_j \ge 0,\ j = 1, 2, ..., n$$

where the decision variable set $\cup_k \{x_k \mid k = 1, 2, ..., K\} = \{x_1, x_2, ..., x_n\} = \{x\}$ and x_K is the control variable vector of the kth-level DM.

The solution process proceeds as follows. The top-level DM provides his or her preferred ranges of f_1 and x_1 to the second-level DM, who solves his or her problem based on the preference information, in terms of membership functions, of the top level. The solution process is essentially the same as that discussed in Section 6.1. The resulting satisfactory solution based on membership functions become additional constraints for the third-level DM. The solution of the third-level DM is presented to the upper levels. If any upper level is not satisfied with this solution, the third-level DM will then solve a new problem with new membership functions until a satisfactory solution is reached. This procedure continues until the kth-level DM, whose solution must satisfy all DMs.

With the required membership functions represented as that of Equations 6.3 and 6.4, the kth-level DM has the following problem:

$$\max_{x_k} f_k(x) \tag{6.23}$$

subject to:

$$\begin{aligned}
& x \in X \\
& \mu_{f1}(f_1(x)) \ge \alpha_1 \text{ and } \mu_{x1}(x_1) \ge \beta_1 \\
& \mu_{f2}(f_2(x)) \ge \alpha_2 \text{ and } \mu_{x2}(x_2) \ge \beta_2 \\
& \cdots \\
& \mu_{f(k-1)}(f_{(k-1)}(x)) \ge \alpha_{(k-1)} \text{ and } \mu_{x(k-1)}(x_{(k-1)}) \ge \beta_{(k-1)} \\
& \alpha_i \in [0, 1],\ i = 1, 2, \ldots, (k-1) \\
& \beta_i \in [0, 1],\ i = 1, 2, \ldots, (k-1)
\end{aligned}$$

Following the same approach as that discussed in Section 6.1, we finally obtain the following expression for a problem with K levels:

$$\max_{x_k} f_k(x) \tag{6.24}$$

subject to:

$$\begin{aligned}
& x \in X \\
& \mu_{f1}(f_1(x)) \ge \alpha_1 \text{ and } \mu_{x1}(x_1) \ge \beta_1 \\
& \mu_{f2}(f_2(x)) \ge \alpha_2 \text{ and } \mu_{x2}(x_2) \ge \beta_2 \\
& \cdots \\
& \mu_{f(K-1)}(f_{(K-1)}(x)) \ge \alpha_{(K-1)} \text{ and } \mu_{x(K-1)}(x_{(K-1)}) \ge \beta_{(k-1)} \\
& \mu_{fK}(f_K(x)) \ge \alpha_K \\
& \alpha_i \in [0, 1],\ i = 1, 2, \ldots, k \\
& \beta_i \in [0, 1],\ i = 1, 2, \ldots, (k-1)
\end{aligned}$$

or

$$\max \lambda \tag{6.25}$$

subject to:

$$\begin{aligned}
&x \in X \\
&\mu_{f1}(f_1(x)) \geq \lambda \text{ and } \mu_{x1}(x_1) \geq \lambda \\
&\mu_{f2}(f_2(x)) \geq \lambda \text{ and } \mu_{x2}(x_2) \geq \lambda \\
&\ldots \\
&\mu_{f(K-1)}(f_{(K-1)}(x)) \geq \lambda \text{ and } \mu_{x(K-1)}(x_{(K-1)}) \geq \lambda \\
&\mu_{fK}(f_K(x)) \geq \lambda \\
&\lambda \in [0, 1]
\end{aligned}$$

where $\lambda = \min\{\alpha_1, \alpha_2, ..., \alpha_{(k-1)}, \alpha_K, \beta_1, \beta_2, \ldots, \beta_{(k-1)}\}$.

Using Equation 6.25, the (K-1) auxiliary problems can be solved. If the solution does not satisfy some of the upper-level DMs, the current lowermost DM must solve the problem again with modified membership functions.

Example 6.3. Consider the following three-level problem which was solved before in Example 3.3 and also by Anandalingdam [1988]:

$$\max_{x1} f_1 = 7x_1 + 3x_2 + 4x_3$$

where x_2 and x_3 solve:

$$\max_{x2} f_2 = x_2$$

where x_3 solves:

$$\max_{x3} f_3 = x_3$$

subject to:

$$\begin{aligned}
&x_1 + x_2 + x_3 \leq 3 \\
&x_1 + x_2 - x_3 \leq 1 \\
&x_1 + x_2 + x_3 \geq 1 \\
&-x_1 + x_2 + x_3 \leq 1 \\
&x_3 \leq 0.5 \\
&x_1, x_2, x_3 \geq 0
\end{aligned}$$

For simplicity, X will be used to represent the above set of constraints.

First, using the constraint set X, the problem is solved individually for each DM. The solutions obtained are $f_1^* = 8.5$ with $x^{1*} = (1.5, 0, 0.5)$; $f_2^* = 1$ with $x^{2*} = (0, 1, 0)$ or $(0.5, 1, 0.5)$; and $f_3^* = 0.5$ with $x^{3*} = (1.5, 0, 0.5)$, $(0.5, 1, 0.5)$, or $(0, 0.5, 0.5)$. The results are also listed in Table 6.2, where the items with * represents the optimum solution. Based on this solution, the limits are set at $f_1' = 3$, $f_2' = 0$ and $f_3' = 0$. Assuming the decision variable x_1 should be around 1.5 with negative and positive-sides tolerances 1.5 and 0, respectively. With $\lambda = \min(\alpha_1, \beta_1, \alpha_2)$, the auxiliary problem for the second level should be:

Table 6.2. Solutions of individual objectives, Example 6.3

Vertex	*f1*	*f2*	(*f3*)	Note
First problem				
A (1.5, 0, 0.5)	8.5*	0	(0.5)	Level 1 optimum
B (0, 1, 0)	3	1*	(0)	Level 2 optimum
C (0.5, 1, 0.5)	4.5	1*	(0.5)	Alternative optimum
D (0.92, 0.58, 0.5)	6.18	0.58	(0.5)	Compromise optimum
Second problem				
E (1.5, 0, 0.5)	8.5	0	0.5*	Level 3 optimum
F (0.5, 1, 0.5)	4.5	1	0.5*	Alternative optimum
G (0, 0.5, 0.5)	-0.5	0.5	0.5*	Alternative optimum
Modified fuzzy range				
	n=1	*n*=2	*n*=3	
fn	(0, 6.18, 8.5)	(0, 0.58, 1)	(0, 0.5, 0.5)	Objective
xn	(0, 0.92, 1.5)	(0, 0.58, 1)	-	Decision variable

$$\max_{x2} f_2 = x_2$$

subject to:

$$\begin{aligned}
&\boldsymbol{x} \in \boldsymbol{X} \\
&(7x_1 + 3x_2 + 4x_3 - 3) / (8.5 - 3) \geq \alpha_1 \\
&x_1 \geq 1.5\beta_1 \\
&x_1 \leq 1.5 \\
&\alpha_1, \beta_1 \in [0, 1]
\end{aligned}$$

or

$$\max \lambda$$

subject to:

$$\begin{aligned}
&\boldsymbol{x} \in \boldsymbol{X} \\
&(7x_1 + 3x_2 + 4x_3 - 3) / (8.5 - 3) \geq \lambda \\
&x_1 \geq 1.5\lambda \\
&x_1 \leq 1.5 \\
&\lambda \in [0, 1]
\end{aligned}$$

Solving this second-level problem, the compromised optimal solution is $\boldsymbol{f} = (f_1, f_2) = (6.18, 0.58)$ with $\boldsymbol{x} = (x_1, x_2, x_3) = (0.92, 0.58, 0.5)$. The satisfaction level for this solution is $\lambda = 0.58$. This information is transferred to the third level. Since the information has been compromised, some modifications are needed to meet the optimum for the whole organization. After modification, the triangular fuzzy numbers for the objective functions and decisions are $f_1 \in [0, 6.18, 8.5]$, $f_2 \in [0, 0.58, 1]$ and $x_1 \in [0, 0.92, 1.5]$, and $x_2 \in [0, 0.58, 1]$ (see Table (6.2)).

Now, the third-level problem can be formulated in the same way as before. Let $\lambda = \min(\alpha_1, \beta_1, \alpha_2, \beta_2, \alpha_3)$ and based on Equations 6.24 and 6.25, we have the auxiliary problem for the third level:

$$\max_{x3} f_3 = x_3$$

subject to:

$$\begin{aligned}
&\boldsymbol{x} \in \boldsymbol{X} \\
&(7x_1 + 3x_2 - 4x_3 - 0) / (6.18 - 0) \geq \alpha_1 \\
&(8.5 - 7x_1 - 3x_2 + 4x_3) / (8.5 - 6.18) \geq \alpha_1 \\
&(\boldsymbol{x}_1 - 0) / (0.92 - 0) \geq \beta_1 \\
&(1.5 - \boldsymbol{x}_1) / (1.5 - 0.92) \geq \beta_1 \\
&(x_2 - 0) / (0.58 - 0) \geq \alpha_2 \\
&(1 - x_2) / (1 - 0.58) \geq \alpha_2 \\
&(x_2 - 0) / (0.58 - 0) \geq \beta_2 \\
&(1 - x_2) / (1 - 0.58) \geq \beta_2 \\
&\alpha_1, \beta_1, \alpha_2, \beta_2 \in [0, 1]
\end{aligned}$$

or

$$\max \lambda$$

subject to:

$$\begin{aligned}
&\boldsymbol{x} \in \boldsymbol{X} \\
&7x_1 + 3x_2 - 4x_3 \geq 6.18\,\lambda \\
&7x_1 - 3x_2 + 4x_3 \leq -2.32\,\lambda + 8.5 \\
&\boldsymbol{x}_1 \geq 0.92\,\lambda \\
&\boldsymbol{x}_1 \leq -0.58\,\lambda + 1.5 \\
&x_2 \geq 0.58\,\lambda \\
&x_2 \leq -0.42\,\lambda + 1 \\
&x_3 \geq 0.5\,\lambda \\
&\lambda \in [0, 1]
\end{aligned}$$

where X is the constraint set of the original problem.

The compromise solution obtained is $\boldsymbol{f} = (f_1, f_2, f_3) = (6.18, 0.58, 0.5)$ with $\boldsymbol{x} = (x_1, x_2, x_3) = (0.92, 0.58, 0.5)$. The satisfaction level is (1, 1, 1). In comparison with the results of Example 3.3, which is $\boldsymbol{f} = (8.5, 0, 0.5)$ with $\boldsymbol{x} = (1.5, 1, 0.5)$ and with satisfaction level of (1, 0, 1), the present solution is much better balanced.

6.4 Fuzzy Multi-level Interactive Decision Making with Multi-followers

Structurally, the most general hierarchy system is the decentralized organization, which has multi-level and with more than one division in each level except the top level. This hierarchical structure is illustrated in Figure 6.2. Mathematically, this structure can be represented by the following set of equations:

$$\max_{x1} f_{11}(\boldsymbol{x}) = \Sigma_j \, c_{11j}{}^T \boldsymbol{x}_j,\ j = 1, 2, \ldots, n \qquad \text{(1}^{\text{st}}\text{ level) (6.26a)}$$

where $\boldsymbol{x}_{21}, \boldsymbol{x}_{22}, \ldots, \boldsymbol{x}_{2p}$ individually solve:

$$\begin{cases} \max_{x21} f_{21}(\boldsymbol{x}) = \Sigma_j \, c_{21j}{}^T \boldsymbol{x}_j,\ j = 1, 2, \ldots, n \\ \ldots \\ \max_{x2p} f_{2p}(\boldsymbol{x}) = \Sigma_j \, c_{2pj}{}^T \boldsymbol{x}_j,\ j = 1, 2, \ldots, n \end{cases} \qquad \text{(2}^{\text{nd}}\text{ level) (6.26b)}$$

where $\boldsymbol{x}_{31}, \boldsymbol{x}_{32}, \ldots, \boldsymbol{x}_{3q}$ individually solve:

$$\begin{cases} \max_{x31} f_{31}(\boldsymbol{x}) = \Sigma_j \, c_{31j}{}^T \boldsymbol{x}_j,\ j = 1, 2, \ldots, n \\ \ldots \\ \max_{x3q} f_{3q}(\boldsymbol{x}) = \Sigma_j \, c_{3qj}{}^T \boldsymbol{x}_j,\ j = 1, 2, \ldots, n \end{cases}$$

$$\text{(3}^{\text{rd}}\text{ level) (6.26c)}$$

where $\boldsymbol{x}_{41}, \boldsymbol{x}_{42}, \ldots, \boldsymbol{x}_{4r}$ individually solve:

$$\begin{cases} \max_{x41} f_{41}(\boldsymbol{x}) = \Sigma_j \, c_{41j}{}^T \boldsymbol{x}_j,\ j = 1, 2, \ldots, n \\ \ldots \\ \max_{x4r} f_{4r}(\boldsymbol{x}) = \Sigma_j \, c_{4rj}{}^T \boldsymbol{x}_j,\ j = 1, 2, \ldots, n \end{cases}$$

$$\text{(4}^{\text{th}}\text{ level) (6.26d)}$$

...

where $\boldsymbol{x}_{K1}, \boldsymbol{x}_{K2}, \ldots, \boldsymbol{x}_{Ks}$ individually solve:

$$\begin{cases} \max_{xk1} f_{K1}(\boldsymbol{x}) = \Sigma_j \, c_{K1j}{}^T \boldsymbol{x}_j,\ j = 1, 2, \\ \ldots, n \\ \ldots \\ \max_{xks} f_{Ks}(\boldsymbol{x}) = \Sigma_j \, c_{Ksj}{}^T \boldsymbol{x}_j,\ j = 1, 2, \\ \ldots, n \end{cases}$$

$$(k^{\text{th}}\text{ level) (6.26e)}$$

subject to:

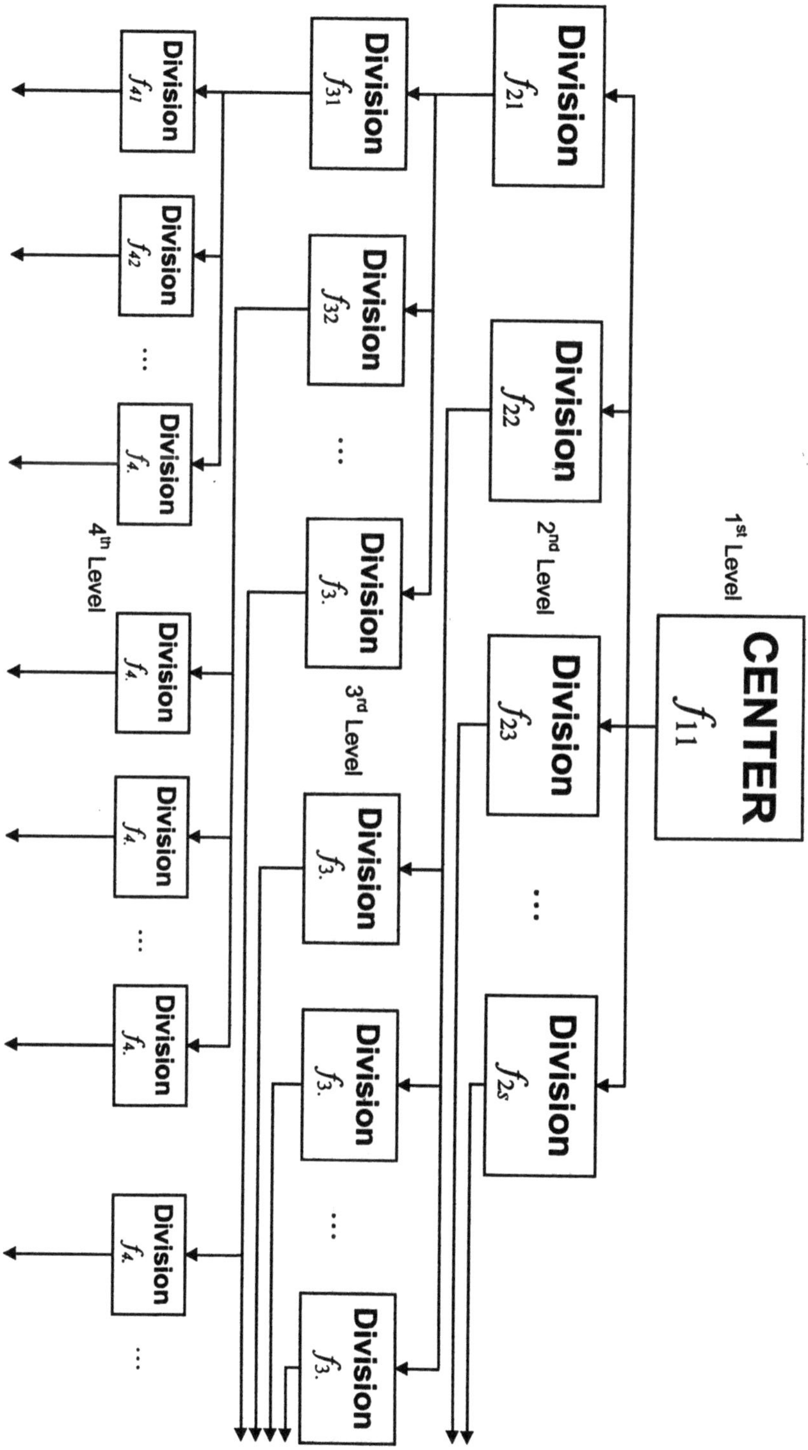

Figure 6.2. Decentralized multi-level hierarchical structure

$$\Sigma_j A_j x_j \leq b, j = 1, 2, \ldots, n$$
$$x_j \geq 0, \ j = 1, 2, \ldots, n$$
$$n = 1 + p \ q \ \ldots \ s$$

where $j = 1, 2, \ldots, n$ represents the decision variable vector of the jth division, and $k = 1, 2, \ldots, K$ represents the kth level. It should be noted that there are p divisions at the second level. For the third level, there are q divisions under each division of the second level. Or, a total of qp divisions at the third level. For example, the third-level divisions $x_{31}, \ldots, x_{3(q1)}$ are under the control of division 21 in the second level and the divisions $x_{3(q1+1)}, \ldots, x_{3(q1+q2)}$ are under the control of division 22 in the second level. For simplicity, we did not indicate all the sub-divisions for the third level. In other words, Equation 6.26c should have repeated p times. Each p repetitions are under the control of one of the divisions in the second level. For the fourth level, there are r divisions under each division of the third level. Or, Equation 6.26d should have repeated q times. Thus, for levels $k = 3, 4, \ldots, K$, not all the sub-divisions are listed in Equation 6.26. We also assumed that, for the kth level, there are s divisions under each division of the $(k\text{-}1)$-th level.

Also for simplicity, the control variables, which are to be maximized, are not indicated or separated from the other variables on the right-hand sides of the maximizing equations in Equation 6.26.

The approaches discussed in the previous two sections can be extended to solve Equation 6.26. However, because of the complexity, it is difficult to propose a general methodology to solve this generalized equation.

Furthermore, because of the constraints at the different levels, there even is the possibility of nonexistence of a feasible solution. If the constraints can be decomposed into individual levels and divisions, then we can decompose the original structure into substructures, which are composed of $(K\text{-}1)$ single upper-level DMs, and many divisions or DMs forming the lower level in the kth level. Then the approaches discussed in the previous sections can be applied in a straightforward manner to solve this problem.

Unfortunately, in many cases the constraints must be considered together for all the DMs in all the levels and divisions. In this case, assuming the bottom levels have feasible spaces, we can start from the bottom levels and solve the problem sequentially up ward. For example, by only considering the bottom three levels, we can treat the problem as a three-level problem. First define $f_{ki}(x)$ as well as x_{ki} as the objective functions and control variable vectors of the kth-level at the ith division, and c_{kij} as the cost coefficient of the decision variable x_j, $j = 1, 2, \ldots, n$, for the ith division of the kth level. Let us assume that there is one division at the top level, p divisions at the second level, and q divisions at the third level. Then, by extending Equations 6.15 and 6.22, the problem can be formulated as:

$$\max_{x1} f_{11}(x) = \Sigma_j c_{11j} x_j \qquad (1^{st} \text{ level}) \quad (6.27)$$

where $x_{21}, x_{22}, \ldots, x_{2p}, x_{31}, x_{32}, \ldots, x_{3q}$ solve:

$$\begin{cases} \max_{x21} f_{21}(\boldsymbol{x}) = \sum_j c_{21j}\, \boldsymbol{x}_j \\ \dots \\ \max_{x2p} f_{2p}(\boldsymbol{x}) = \sum_j c_{2pj}\, \boldsymbol{x}_j \end{cases} \qquad (2^{nd}\text{ level})$$

where $\boldsymbol{x}_{31}, \boldsymbol{x}_{32}, \dots, \boldsymbol{x}_{3q}$ solve:

$$\begin{cases} \max_{x31} f_{31}(\boldsymbol{x}) = \sum_j c_{31j}\, \boldsymbol{x}_j \\ \dots \\ \max_{x3q} f_{3q}(\boldsymbol{x}) = \sum_j c_{3qj}\, \boldsymbol{x}_j \end{cases} \qquad (3^{rd}\text{ level})$$

subject to:

$$\sum_{k,j} \boldsymbol{A}_{kj}\, \boldsymbol{x}_{kj} \le \boldsymbol{b}, \quad k = 1, 2, \dots, K$$
$$\boldsymbol{x}_{kj} \ge 0, \quad j = 1, 2, \dots, n$$

where $j = 1, 2, \dots, n$ represents the jth decision variable. The decision variable set $\cup_{k,i} \{\boldsymbol{x}_{kj} \mid \forall\, i \text{ and } k\} = \{\boldsymbol{x}_1, \boldsymbol{x}_2, \dots, \boldsymbol{x}_n\}$. For the second level, we have the following auxiliary problem:

$$\max \lambda \qquad (6.28)$$

subject to:

$$\boldsymbol{x} \in \boldsymbol{X}$$
$$\mu_{f11}(f_{11}(\boldsymbol{x})) \ge \lambda \text{ and } \mu_{x11}(\boldsymbol{x}_{11}) \ge \lambda$$
$$\mu_{f2i}(f_{2I}(\boldsymbol{x})) \ge \lambda, \quad i = 1, 2, \dots, p$$
$$\lambda \in [0, 1]$$

where $\boldsymbol{X}$ represents the original constraint of the problem. This expression is similar to Equation 6.25. Using the satisfactory solutions of the first and the second levels in the forms of membership functions, the third-level auxiliary problem becomes:

$$\max \lambda \qquad (6.29)$$

subject to:

$$\boldsymbol{x} \in \boldsymbol{X}$$
$$\mu_{f11}(f_{11}(\boldsymbol{x})) \ge \lambda \text{ and } \mu_{x11}(\boldsymbol{x}_{11}) \ge \lambda$$
$$\mu_{f21}(f_{21}(\boldsymbol{x})) \ge \lambda \text{ and } \mu_{x21}(\boldsymbol{x}_{21}) \ge \lambda$$
$$\dots$$
$$\mu_{f2p}(f_{2p}(\boldsymbol{x})) \ge \lambda \text{ and } \mu_{x2p}(\boldsymbol{x}_{2p}) \ge \lambda$$
$$\mu_{f3I}(f_{3i}(\boldsymbol{x})) \ge \lambda, \quad i = 1, 2, \dots, q$$
$$\lambda \in [0, 1]$$

Compromise solution for the three level with multi-followers can be obtained by solving the above equation. If the solution does not satisfy some of the DMs, an interactive process as that discussed earlier can be used.

If the constraint set or feasible domain is non-separable, DMs in the same level need to make their decisions simultaneously. Even under this condition, the problem is not necessarily reduced to a single-person decision-making problem [Bard 1983a]. In fact, this problem can be treated as a multiple-objective decision-making problem. However, if parallel computer system is used, this multi-person, decision-making processes can be carried out as a one-person decision-making problem.

Although much more computation is involved, the general hierarchical structure shown in Figure 6.2 can be solved by using the fuzzy interactive scheme represented by Equation 6.27-6.29. The process is essentially an interactive trial-and-error approach. Depending on how important is the results, various re-adjustments and re-computation can be carried out by using fuzzy membership functions.

6.5 Discussions

The proposed decision-making process proceeds from top to bottom in a natural and straightforward manner. Preference information is delivered from upper levels to lower levels sequentially, and the lower-level DM solves his or her problem under the restrictions of the upper-level DM's requirements. Since both decision variables and objective functions are considered, the proposed satisfactory solution should be more practical and reasonable. Since the solution search is based on the change of membership function instead of vertex enumeration, even a large-scale problem can be solved with reasonable amount of computation. Furthermore, for non-linear programming problem, the proposed approach is still applicable although the amount of computation can increase tremendously depending on the particular problem.

The fuzzy membership functions in our procedure are assumed to be linearly increasing or linearly decreasing. Obviously, non-linear membership function can also be used, which is important in neural network applications. Some of the non-linear membership functions are exponential or the Gaussian function.

Since different membership function and operation provide different satisfactory solution, it is important to explore various functions and operators, as well as allow the DM to change functional forms in the interactive processes. In this respect, experimental work for investigating human behavior could provide the foundation for the formulation of different membership functions and gain more understanding of the human decision-making process.

Finally, we would like to mention that input data is often imprecise or fuzzy in practice for a wide variety of hierarchical optimization problems, such as defense problems, economical analyses, government regulations, organizational management and so on. Developing methodologies and new concepts, even embedded multiple objectives in each level, for solving fuzzy, multi-level programming problems is a practical and interesting direction for future study.

Chapter 7

Aggregation of Fuzzy Systems in Multi-level Decisions

Although the various approaches proposed for fuzzy multi-level programming in the previous chapter are very powerful, it is only a start in this fruitful direction. Further developments are needed to consider other aspects and thus to make the actual solution process more practical in the sense of solving more realistic problems with reduced computation. Two of the most important aspects which need to be considered are the problem of aggregation of the fuzzy representation and the representation of systems whose description is vague and not well defined. The first aspect is concerned with the basic problem of fuzzy combination, where, in the previous chapter, only the non-compensatory *min* is used. The second aspect is concerned with the handling of vague systems. Notice that the problem representing the system at each division or each level in the previous chapter is crisp, not fuzzy. The fuzziness comes from the mutual interaction or mutual competition of the different decision-makers in each level or each division. In this chapter, fuzzy systems represented by fuzzy parameters or fuzzy coefficients, not just fuzzy objectives and fuzzy decisions, will be considered.

Fuzzy information aggregation. Fuzzy number operation not only forms the basic framework of fuzzy set theory but also plays a central role in fuzzy decision making. The basic problem in fuzzy number operation is that fuzzy numbers are sets, not crisp numbers, and these sets frequently are partial order, not linear order. Thus fuzzy numbers can be combined or aggregated in different ways depending on the usage or the purpose of the combination. In set theoretic approach, fuzzy number operations are discussed in terms of the logical conjunction *and* and the logical disjunction *or*. Bellman and Zadeh [1970], in a seminal paper, interpreted the *and* and the *or* as the *max* and the *min* operators, respectively, in decision making. These interpretations of the connectives *and* and *or* are generally used in the early stages of fuzzy set theory development. Bellman and Giertz [1973] and Fung and Fu [1975], among others, have examined the rational assumptions as well as the axioms of the generalized *max* and *min* operators. However, the aggregation of partial ordered fuzzy numbers is not straightforward and different emphasis or different interpretation may be needed depending on the application. Many other interpretations for the connectives *and* and *or* have been proposed by various investigators. For a detailed discussion, the reader can consult the literature [Klir and Folger 1988, Zhu 1992, Zimmermann 1991].

Table 7.1. Dual pairs of common operators

A. Compensatory Minimum Operators

(A.1) Algebraic product: $\mu = \alpha \cdot \beta$;
(A.2) Bounded product: $\mu = \max(0, \alpha \cdot \beta + 1)$;
(A.3) Hamacher's min operator:
$\mu = \alpha \cdot \beta / [\Gamma + (1-\Gamma)(\alpha+\beta-1)], \Gamma \in [0,1]$;
(A.4) Yager's min operator:
$\mu = 1 - \min\{1, [(1-\alpha)^q + (1-\beta)^q]^{1/q}\}, q \geq 1$;
(A.5) Dubois and Prade's min product:
$\mu = \alpha \cdot \beta / [\max(\alpha, \beta, \gamma)], \gamma \in [0,1]$;
(A.6) Werners' "fuzzy and" operator:
$\mu = \gamma \cdot \min(\alpha, \beta) + (1-\gamma)(\alpha+\beta)/2, \gamma \in [0,1]$;
(A.7) Einstein product: $\mu = (\alpha \cdot \beta) / [2-(\alpha+\beta-\alpha \cdot \beta)]$.

B. Compensatory Maximum Operators

(B.1) Algebraic sum: $\mu = \alpha + \beta - \alpha \cdot \beta$;
(B.2) Bounded sum: $\mu = \min(1, \alpha+\beta)$;
(B.3) Hamacher's max operator:
$\mu = [(1-\Gamma) \cdot \alpha \cdot \beta + \Gamma(\alpha+\beta)] / (\Gamma+\alpha \cdot \beta), \Gamma \in [0,1]$;
(B.4) Yager's max operator: $\mu = \min\{1, (\alpha^q+\beta^q)^{1/q}\}, q \geq 1$;
(B.5) Dubois and Prade's max product:
$\mu = [(\alpha+\beta-\alpha \cdot \beta) - \min(1-\gamma, \alpha, \beta)] / [\max(\gamma, 1-\alpha, 1-\beta)]$,
$\gamma \in [0,1]$;
(B.6) Werners' "fuzzy or" operator:
$\mu = \gamma \cdot \max(\alpha, \beta) + (1-\gamma)(\alpha+\beta)/2, \gamma \in [0,1]$;
(B.7) Einstein sum: $\mu = (\alpha+\beta) / (1+\alpha \cdot \beta)$.

As has been mentioned in the two previous chapters, the most important aspect of these interpretations is whether or not the operator is compensatory, or, in other words, whether or not trade-off between low value and high value is allowed. Since in practice, managerial decisions almost always allow some degree of compensation, compensatory operators are obviously desirable in decision making. Some of the compensatory operators are listed in Table 7.1 for problems with two operands, α and β. The notations Γ and γ in Table 7.1 are weighting parameters. Beside the compensatory behavior, we must also consider whether the operator is easy to use and not introduces non-linear and other computational difficult elements. For example, Zimmermann and Zysno [1980] have shown that the Hamacher's γ-operator, which is listed as Item 3 in Table 7.1 predicts human judgment quite well, but it is a fairly complex expression, which will introduce non-linear elements and causes considerable computational difficulty. Based on the γ-operator, Werners [1988] proposed the fuzzy *and* and the fuzzy *or* operators, which add a compensatory element (see Item 6 in Table 7.1) and are easy to handle. Reasonably consistent results have been obtained by using these operators. The Werners interpretation will be adopted in the following.

7.1 Compensation in Bi-level Decisions

The fuzzy bi-level *programming* problem obtained in Chapter 6 and was represented by Equation 6.13 is reproduced in the following:

$$\max \lambda \tag{7.1}$$

subject to:

$$\begin{aligned}
&\boldsymbol{A}_1 \boldsymbol{x}_1 + \boldsymbol{A}_2 \boldsymbol{x}_2 \leq \boldsymbol{b} \\
&[f_1(\boldsymbol{x}) - f_1'] / [f_1^{L} - f_1'] \geq \lambda \\
&[(\boldsymbol{x}_1^{U} + \boldsymbol{p}_1) - \boldsymbol{x}_1)] / \boldsymbol{p}_1 \geq \lambda \\
&[\boldsymbol{x}_1 - (\boldsymbol{x}_1^{U} - \boldsymbol{p}_1)] / \boldsymbol{p}_1 \geq \lambda \\
&[f_2(\boldsymbol{x}) - f_2'] / [f_2^{L} - f_2'] \geq \lambda \\
&\lambda \in [0, 1] \\
&\boldsymbol{x}_1, \boldsymbol{x}_2 \geq 0
\end{aligned}$$

where $\boldsymbol{x}_1^{U}$ is the preferred value of $\boldsymbol{x}_1$ and $\boldsymbol{p}_1$ is the two-side tolerance for $\boldsymbol{x}_1$, and f_1^{T} and f_1' is the upper and lower bounds of f_1, respectively.

Recall that λ was obtained by using the *min* operator as follows:

$$\lambda = \min (\alpha, \beta, \gamma) \tag{7.2}$$

where α, β and γ are the minimum acceptable degrees of satisfaction, respectively, for the objective function $f_1(x)$, the decision x_1, and the objective function $f_2(x)$. Instead of using the *min* operator, which is non-compensatory, we wish to use the operator proposed by Werners [1988], which is listed in Table 7.1 and is reproduced in the following:

$$\mu_{\text{and}} = \gamma \min_i (\mu_i) + (1-\gamma) (\Sigma_i \mu_i) / m \tag{7.3a}$$

$$\mu_{\text{or}} = \gamma \max_i (\mu_i) + (1-\gamma) (\Sigma_i \mu_i) / m \tag{7.3b}$$

where the membership functions, μ_i, $0 \leq \mu_i \leq 1$, $i = 1, 2, ..., m$, and $\gamma \in [0, 1]$ is the degree of compensation or the degree of approximation for the logical connectives *and* and *or*.

These operators are the convex combinations of the *min* or the *max* operators with the arithmetical mean. They furnish the compensatory effect without introducing too much computational requirements. Let $\lambda = \min_i (\mu_i)$, then, the connective *and* can be further simplified:

$$
\begin{aligned}
\mu_{\text{and}} &= \gamma\,\lambda + (1-\gamma)\,(\Sigma_i\,\mu_i)\,/\,m = \gamma\,\lambda + (1-\gamma)\,\Sigma_i\,((\mu_i - \lambda) + \lambda)\,/\,m \\
&= \gamma\,\lambda + (1-\gamma)\,m\,\lambda/\,m + (1-\gamma)\,\Sigma_i\,(\mu_i - \lambda)\,/\,m \\
&= \lambda\ + (1-\gamma)\,\Sigma_i\,(\mu_i - \lambda)\,/\,m
\end{aligned}
$$

where $0 \le \mu_i \le 1$, $i = 1, 2, \ldots, m$, $0 \le \lambda \le 1$, $0 \le \gamma \le 1$. Let $\lambda_i = \mu_i - \lambda$, the above expression is further reduced to:

$$\mu_{\text{and}} = \lambda + (1-\gamma)\,\Sigma_i\,\lambda_i\,/\,m \tag{7.4}$$

Equation 7.4 gives a compromised membership function. If we wish to obtain a crisp answer, we would like to obtain the maximum of this membership function. Thus, we have

$$\max\,\mu_{\text{and}} = \lambda + (1-\gamma)\,\Sigma_i\,\lambda_i\,/\,m \tag{7.5}$$

Thus, a problem for the aggregation of m membership functions with given constraints $\boldsymbol{x} \in \boldsymbol{X}$ can be formulated as:

$$\max\,\mu_{\text{and}} = \lambda + (1-\gamma)\,\Sigma_i\,\lambda_i\,/\,m \tag{7.6}$$

subject to:

$$
\begin{aligned}
&\boldsymbol{x} \in \boldsymbol{X} \\
&\mu_i \ge \lambda + \lambda_i \\
&\lambda + \lambda_i \le 1 \\
&\gamma \in [0, 1] \\
&\lambda,\ \mu_i,\ \lambda_i \in [0, 1]
\end{aligned}
$$

with $i = 1, 2, \ldots, m$ and $\boldsymbol{X}$ is a crisp constraint set.

For the bi-level programming problem, we need to aggregate between the three minimum acceptable levels, α, β and γ. Thus, m = 3. Combining Equation 7.5 with the constraints in Equation 7.1 for the original problem, we finally have the desired equation:

$$\max\,\mu_{\text{and}} = \lambda + (1-\gamma)\,(\lambda_1 + \lambda_2 + \lambda_3)\,/\,3 \tag{7.7}$$

subject to:

$$
\begin{aligned}
&\boldsymbol{A}_1\,\boldsymbol{x}_1 + \boldsymbol{A}_2\,\boldsymbol{x}_2 \le \boldsymbol{b} \\
&\boldsymbol{x}_1, \boldsymbol{x}_2 \ge 0 \\
&\mu_{f1}\,(f_1(\boldsymbol{x})) = [f_1(\boldsymbol{x}) - f_1']\,/\,[f_1^{\text{L}} - f_1'] \ge (\lambda + \lambda_1) \\
&[(\boldsymbol{x}_1^{\text{U}} + \boldsymbol{p}_1) - \boldsymbol{x}_1]\,/\,\boldsymbol{p}_1 \ge (\lambda + \lambda_2) \\
&[\boldsymbol{x}_1 - (\boldsymbol{x}_1^{\text{U}} - \boldsymbol{p}_1)\,]\,/\,\boldsymbol{p}_1 \ge (\lambda + \lambda_2) \\
&\mu_{f2}\,(f_2(\boldsymbol{x})) = [f_2(\boldsymbol{x}) - f_2']\,/\,[f_2^{\text{L}} - f_2'] \ge (\lambda + \lambda_3\,) \\
&\lambda + \lambda_i \le 1,\ i = 1, 2, 3 \\
&\gamma \in [0, 1]
\end{aligned}
$$

$$\lambda, \lambda_1, \lambda_2, \lambda_3 \in [0, 1]$$

where $\boldsymbol{p}_1$ is the two-sided tolerance for the decision vector $\boldsymbol{x}_1$ and γ is the grade of compensation.

Although we specified a minimum compensation level, we could also allow a compensation level within the interval [0,1]. Since this is a linear problem, it can be solved easily. The following example resolved the problem solved in Chapter 6.

Example 7.1. Consider the problem solved in Example 6.1, where the upper and lower solutions obtained were; $x^U = (x_1^U, x_2^U) = (7.5, 1.5)$ for $f_1^U = 13.5$ and $\boldsymbol{x}^L = (x_1^L, x_2^L) = (3, 9)$ for $f_2^L = 21$, which were used as reference. Now, let $f_1^U = 13.5$ and $f_1' = 0$ instead of -3, and $f_2' = 10.5$. Further assume that the upper-level DM's control decision x_1 is around 7.5 with plus or minus tolerances 4.5 and 0.5, respectively. With these assumed values, we can establish the membership functions for μ_{x1}, μ_{f1}, and μ_{f2}. The lower-level DM then solves the following problem, which represents an auxiliary model for the given bi-level problem in Example 6.1.

$$\max \lambda$$

subject to:

$$\begin{aligned}
&\boldsymbol{x} \in X \\
&\mu_{f1}(f_1(\boldsymbol{x})) = f_1 / (13.5 - 0) \geq \lambda \\
&\mu_{x1}(x_1) = (x_1 - 4.5) / (7.5 - 4.5) \geq \lambda \\
&\mu_{x1}(x_1) = (8 - x_1) / (8 - 7.5) \geq \lambda \\
&\mu_{f2}(f_2(\boldsymbol{x})) = (f_2 - 10.5) / (21 - 10.5) \geq \lambda \\
&\lambda \in [0, 1]
\end{aligned}$$

where $\boldsymbol{X}$ represents the original crisp constraint set.

Now, the corresponding compensatory formulation for the bi-level problem in Example 6.1 is:

$$\max \mu_{\text{and}} = \lambda + (1-\gamma)(\lambda_1 + \lambda_2 + \lambda_3) / 3$$

subject to:

$$\begin{aligned}
&\boldsymbol{x} \in X \\
&(f_1 - 0) / (13.5 - 0) \geq (\lambda + \lambda_1) \\
&(x_1 - 4.5) / (7.5 - 4.5) \geq (\lambda + \lambda_2) \\
&(8 - x_1) / (8 - 7.5) \geq (\lambda + \lambda_2) \\
&(f_2 - 10.5) / (21 - 10.5) \geq (\lambda + \lambda_3) \\
&\lambda, \lambda_1, \lambda_2, \lambda_3 \in [0, 1]
\end{aligned}$$

where $\gamma \in [0, 1]$ represents the degree of compensation.

This compensatory formulation can be solved by any linear programming code. The compromise solution is $f^* = (f_1^*, f_2^*) = (9.28, 17.72)$ at $x^* = (x_1^*, x_2^*) = (7.26, 5.23)$, with a total degree of satisfaction $\mu_{and} = 0.73$ at $\gamma = 0.5$. Compared to the solution obtained in Chapter 6, where $\lambda = 0.69$ with the same objective function and decision values, the degree of satisfaction has been improved due to the use of better aggregation operator. To investigate the effect of compensation in more detail, 11 different cases with γ varies from 0 to 1.0 were solved and the results are listed in Table 7.2. The value of μ_{and} varies within a range from 0.69 to 0.78 and λ remains at 0.69, except at γ =0, where $\lambda = 0.57$.

Table 7.2. Results of Example 7.1

γ	μ_{and}	x_1, x_2
0.0	0.78	7.5, 4.5
0.1	0.77	7.26, 5.23
0.2	0.76	7.26, 5.23
0.3	0.75	7.26, 5.23
0.4	0.74	7,26, 5.23
0.5	0.73	7.26, 5.23
0.6	0.72	7,26, 5.23
0.7	0.71	7.26, 5.23
0.8	0.70	7.26, 5.23
0.9	0.70	7.26, 5.23
1.0	0.69	7.26, 5.23

7.2 Compensation in Multiple-level Problems

Based on the discussions in Chapter 6 and the use of the *min* operator, a three-level programming problem can be transformed into the following auxiliary form:

$$\max \lambda \tag{7.8}$$

subject to:

$$A_1 x_1 + A_2 x_2 + A_3 x_3 \le b$$
$$x_1, x_2, x_3 \ge 0$$
$$\mu_{f1}(f_1(x)) = [f_1(x) - f_1']/[f_1^L - f_1'] \ge \lambda$$
$$[(x_1^U + p_1) - x_1]/p_1 \ge \lambda$$
$$[x_1 - (x_1^U - p_1)]/p_1 \ge \lambda$$
$$\mu_{f2}(f_2(x)) = [f_2(x) - f_2']/[f_2^L - f_2'] \ge \lambda$$
$$[(x_2^U + p_2) - x_2]/p_2 \ge \lambda$$
$$[x_2 - (x_2^U - p_2)]/p_2 \ge \lambda$$
$$\mu_{f3}(f_3(x)) = [f_3(x) - f_3']/[f_3^L - f_3'] \ge \lambda$$

where x_1^U is the preferred value of x_1, p_1 and p_2 are the two-sided tolerances for x_1 and x_2, respectively, x_2^U is the preferred value for x_2, f_1^L and f_1' are the upper and lower bounds for f_1, respectively.

Replacing the *min* in Equation 7.8 by the Werners operator, Equation 7.8 can be transformed into the following auxiliary model for the tri-level problem:

$$\max \mu_{and} = \lambda + (1-\gamma)(\lambda_1 + \lambda_2 + \lambda_3 + \lambda_4 + \lambda_5) / 5 \tag{7.9}$$

subject to:

$$\begin{aligned}
& A_1 x_1 + A_2 x_2 + A_3 x_3 \le b \\
& x_1, x_2, x_3 \ge 0 \\
& \mu_{f1}(f_1(x)) = [f_1(x) - f_1'] / [f_1^L - f_1'] \ge (\lambda + \lambda_1) \\
& [(x_1^U + p_1) - x_1] / p_1 \ge (\lambda + \lambda_2) \\
& [x_1 - (x_1^U - p_1)] / p_1 \ge (\lambda + \lambda_2) \\
& \mu_{f2}(f_2(x)) = [f_2(x) - f_2'] / [f_2^L - f_2'] \ge (\lambda + \lambda_3) \\
& [(x_2^U + p_2) - x_2] / p_2 \ge (\lambda + \lambda_4) \\
& [x_2 - (x_2^U - p_2)] / p_2 \ge (\lambda + \lambda_4) \\
& \mu_{f3}(f_3(x)) = [f_3(x) - f_3'] / [f_3^L - f_3'] \ge (\lambda + \lambda_5) \\
& \lambda + \lambda_i \le 1, \ i = 1, 2, 3, 4, 5 \\
& \lambda, \lambda_1, \lambda_2, \lambda_3, \lambda_4, \lambda_5 \in [0, 1]
\end{aligned}$$

where γ is the degree of compensation with $\gamma \in [0.1]$.

The supervised search procedure will be defined in the same way as in Chapter 6, that is, from the first level to the second level and then to the three levels. If the solution of Equation 7.9 does not satisfy the upper-level, modification of the membership functions will be needed with new preference information. Furthermore, it is noted that if there is a constraint, which makes the problem infeasible, this constraint should be examined carefully and may be allow some tolerance so that the original problem becomes feasible.

Example 7.2. Consider the following tri-level problem, which was solved by Anandalingdam [1988] and also in Example 6.3.

$$\max_{x1} f_1 = 7x_1 + 3x_2 - 4x_3$$

where x_2 and x_3 solve:

$$\max_{x2} f_2 = x_2$$

where x_3 solves:

$$\max_{x3} f_3 = x_3$$

subject to:

$$x_1 + x_2 + x_3 \le 3$$

$$x_1 + x_2 - x_3 \le 1$$
$$x_1 + x_2 + x_3 \ge 1$$
$$-x_1 + x_2 + x_3 \le 1$$
$$x_3 \le 0.5$$
$$x_1, x_2, x_3 \ge 0$$

This problem was first solved for each individual decision-maker. The solution was (see also Problem 6.3): $f_1^{1T} = 8.5$ at $x_1^{1U} = (1.5, 0, 0.5)$, $f_2^{1T} = 1$ at $x_2^{1U} = (0, 1, 0)$ or $(0, 1, 0.5)$, and $f_3^{1T} = 0.5$ at $x_3^{1U} = (1.5, 0, 0.5)$, $(0.5, 1, 0.5)$, or $(0, 0.5, 0.5)$. In addition, $f_1' = 0$ instead of -0.5 (only positive value is meaningful), $f_2' = f_3' = 0$. Also, assume that control decision x_1 is around 1.5 with a tolerance plus and minus 1.5. The compensatory auxiliary problem of the second level becomes:

$$\max \lambda + (1-\gamma)(\lambda_1 + \lambda_2 + \lambda_3) / 3$$

subject to:

$$x \in X$$
$$7x_1 + 3x_2 - 4x_3 \ge 8.5(\lambda + \lambda_1)$$
$$x_1 \ge 1.5(\lambda + \lambda_{2,})$$
$$x_1 + 1.5(\lambda + \lambda_2) \le 3$$
$$f_2 = x_2 \ge (\lambda + \lambda_3)$$
$$\lambda + \lambda_i \le 1, \; i = 1, 2, 3$$
$$\lambda, \lambda_1, \lambda_2, \lambda_3 \in [0, 1]$$

where X represents the constraint set of the original problem and γ is the degree of compensation with $\gamma \in [0,1]$.

The compromise solutions obtained for the first- and second-level problems are:

$$f^{2T} = (f_1^{2T}, f_2^{2T}) = (6.1, 0.6), \; x^{2U} = (x_1^{2U}, x_2^{2U}, x_3^{2U}) = (0.9, 0.6, 0.5),$$
$$\gamma = 0.5$$

with a degree of satisfaction $\mu_{and} = 0.62$. Assume that the top two DMs are satisfied with this solution, we then go to the third level of this problem. Since the information has been compromised, some modifications are needed to obtain the optimum of the overall organization. Modify the intervals of decision variables and the goals as:

$$x_1 = [0, 0.9, 1.8], x_2 = [0, 0.6, 2], f_1 = [0, 6.1), \; f_2 = [0, 0.6)$$
$$\text{and } f_3 = [0, 0.5).$$

The auxiliary problem of the third level is thus formulated as the following crisp problem:

$$\max \mu_{and} = \lambda + (1-\gamma)(\lambda_1 + \lambda_2 + \lambda_3 + \lambda_4 + \lambda_5) / 5$$

subject to:

$$\begin{aligned}
&\boldsymbol{x} \in \boldsymbol{X} \\
&7\,x_1 + 3\,x_2 - 4\,x_1 \geq 6.1\ (\lambda + \lambda_1) \\
&x_1 \geq 0.9\ (\lambda + \lambda_2) \\
&x_1 + 0.9\ (\lambda + \lambda_2) \leq 1.8 \\
&f_2 = x_2 \geq 0.6\ (\lambda + \lambda_3) \\
&x_2 + (\lambda + \lambda_4) \leq 2 \\
&f_3 = x_3 \geq 0.5\ (\lambda + \lambda_5) \\
&\lambda + \lambda_i \leq 1,\ \ i = 1, 2, 3, 4, 5 \\
&\lambda, \lambda_1, \lambda_2, \lambda_3, \lambda_4, \lambda_5 \in [0, 1]
\end{aligned}$$

where $\boldsymbol{X}$ is the constraint set of the original problem and γ is the degree of compensation with $\gamma \in [0,1]$.

After solving the third-level auxiliary problem, the compensatory solution for the whole system is:

$$\boldsymbol{f}^* = (f_1^*, f_2^*, f_3^*) = (6.1, 0.6, 0.5), \boldsymbol{x}^* = (x_1^*, x_2^*, x_3^*) = (0.9, 0.6, 0.5),$$
$$\gamma = 0.5$$

with a satisfaction level 1.0, that is, complete satisfaction for all the DMs. The crisp solution is $\boldsymbol{f}^* = (4.5, 1, 0.5)$ at $\boldsymbol{x}^* = (0.5, 1, 0.5)$ with satisfaction (1.0, 0, 1.0).

Obviously, the above procedure can be directly extended to the kth-level programming problem, with $k \geq 4$. In this case, we will need to solve k-1 auxiliary problems. However, if the decision-maker is not satisfied, some interactive process will be needed to modify the degree of satisfaction

7.3 Bi-level Decentralized Problem with Equally Important Objectives

We have developed a compensatory fuzzy approach for efficiently solving multi-level mathematical programming problem. In this and the next two sections, we shall extend the procedure to hierarchy decentralized problems. These planning problems include bi-level decentralized programming problem, three-level decentralized programming problem, and, more generally, multi-level decentralized programming problem. The decentralized problem comprises of a decision center at the top level, several divisions at each of the lower levels. This is the generalized multi-level hierarchy structure of Anandalingam's decentralized system [1988]. However, this generalized structure is too complex to be managed both in theory and in application. Furthermore, there is no explicit relationship among the various decision units at the same level and thus it is also difficult to investigate the problem from the standpoint of the Stackelberg game. As a result, very few investigations in this area have been carried out. As we have discussed in Chapter 6, the proposed fuzzy approach can solved these large decentralized problems efficiently. Now, we would like to examine how compensatory operators can be incorporated and thus extend the compensatory approach to this decentralized system

Using the *min* operator, the auxiliary equations for the bi-level problem with s followers were obtained in Equation 6.20, which are repeated in the following:

$$\max \lambda \tag{7.10}$$

subject to:

$$\begin{aligned}
&\boldsymbol{x} \in \boldsymbol{X}\\
&\mu_{f1}(f_1(\boldsymbol{x})) = [f_1(\boldsymbol{x}) - f_1'] / [f_1^{\mathrm{L}} - f_1'] \geq \lambda\\
&[(\boldsymbol{x}_1^{\mathrm{T}} + \boldsymbol{p}_1) - \boldsymbol{x}_1] / \boldsymbol{p}_1 \geq \lambda\\
&[\boldsymbol{x}_1 - (\boldsymbol{x}_1^{\mathrm{T}} - \boldsymbol{p}_1)] / \boldsymbol{p}_1 \geq \lambda\\
&\mu_{f2i}(f_{2i}(\boldsymbol{x})) = [f_{2i}(\boldsymbol{x}) - f_{2i}'] / [f_{2i}^{\mathrm{L}} - f_{2i}'] \geq \lambda, \quad i = 1, 2, \ldots, s\\
&\lambda \in [0, 1]
\end{aligned}$$

where $\boldsymbol{X}$ represents the original constraint set, $\boldsymbol{x}_1^{\mathrm{U}}$ is the preferred value of $\boldsymbol{x}_1$ and $\boldsymbol{p}_1$ represents the two-sided tolerance of $\boldsymbol{x}_1$, and f_1^{L} and f_1' is the upper and lower bounds of f_1, respectively.

Using Werners compensatory operator, the above expression can be transformed into the following compensatory model:

$$\max \mu_{\text{and}} = \lambda + (1-\gamma)\,(\lambda_1 + \lambda_2 + \ldots + \lambda_{s+2}) / (s+2) \tag{7.11}$$

subject to:

$$\begin{aligned}
&\boldsymbol{x} \in \boldsymbol{X}\\
&\mu_{f1}(f_1(\boldsymbol{x})) = [f_1(\boldsymbol{x}) - f_1'] / [f_1^{\mathrm{L}} - f_1'] \geq \lambda + \lambda_1\\
&[(\boldsymbol{x}_1^{\mathrm{T}} + \boldsymbol{p}_1) - \boldsymbol{x}_1] / \boldsymbol{p}_1 \geq (\lambda + \lambda_2)\\
&[\boldsymbol{x}_1 - (\boldsymbol{x}_1^{\mathrm{T}} - \boldsymbol{p}_1)] / \boldsymbol{p}_1 \geq (\lambda + \lambda_2)\\
&\mu_{f21}(f_{21}(\boldsymbol{x})) = [f_{21}(\boldsymbol{x}) - f_{21}'] / [f_{21}^{\mathrm{L}} - f_{21}'] \geq (\lambda + \lambda_3)\\
&\ldots\\
&\mu_{f2s}(f_{2s}(\boldsymbol{x})) = [f_{2s}(\boldsymbol{x}) - f_{2s}'] / [f_{2s}^{\mathrm{L}} - f_{2s}'] \geq (\lambda + \lambda_{s+2})\\
&\lambda + \lambda_i \leq 1, \; i = 1, 2, \ldots, s+2\\
&\lambda, \lambda_1, \lambda_2, \ldots, \lambda_{s+2} \in [0, 1]
\end{aligned}$$

with $i = 1, 2, \ldots, s$ and γ is the degree of compensation.

Example 7.3. Consider the bi-level problem solved in Example 6.2, which was also solved by Anandalingdam [1988].

The individual solutions of the DMs in the top and bottom levels are: $f_{11}^{\mathrm{T}} = 35$ at $\boldsymbol{x}_{11}^{\mathrm{U}} = (x_1^{\mathrm{U}}, y_1^{\mathrm{U}}, y_2^{\mathrm{U}}, y_3^{\mathrm{U}}) = (10, 0, 10, 0)$ or $(5, 5, 10, 3)$, $f_{21}^{\mathrm{T}} = 30$ at $\boldsymbol{x}_{21}^{\mathrm{U}} = (0, 10, 0, 0)$, $f_{22}^{\mathrm{T}} = 30$ at $\boldsymbol{x}_{22}^{\mathrm{U}} = (0, 0, 10, 0)$ and $f_{23}^{\mathrm{T}} = 30$ at $\boldsymbol{x}_{23}^{\mathrm{U}} = (0, 0, 0, 10)$. Since negative values are not allowed, let $f_{11}' = f_{21}' = f_{22}' = f_{23}' = 0$. Assume that for x_1 the upper-level DM is satisfied for any value between 5 and 10, with negative side tolerance of 5 and positive tolerance of 0. Based on these assumptions, the membership functions can be established. Using Equation 7.11, we obtain:

$$\max \mu_{and} = \lambda + (1-\gamma)(\lambda_1 + \lambda_2 + \lambda_3 + \lambda_4 + \lambda_5)/5$$

subject to:

$$\begin{aligned}
& \boldsymbol{x} \in \boldsymbol{X} \\
& \mu_{f11}(f_{11}(\boldsymbol{x})) = [f_{11}(\boldsymbol{x}) - f_{11}'] / [f_{11}^{L} - f_{11}'] \\
& \qquad = (x_1 + y_1 + 2y_2 + y_3)/35 \geq \lambda + \lambda_1 \\
& [x_1 - (x_1^{U} - p_1)] / p_1 = x_1/5 \geq (\lambda + \lambda_2) \\
& x_1 \leq 10 \\
& \mu_{f21}(f_{21}(\boldsymbol{x})) = [f_{21}(\boldsymbol{x}) - f_{21}'] / [f_{21}^{L} - f_{21}'] \\
& \qquad = (-x_1 + 3y_1 - 2y_2 - y_3)/30 \geq (\lambda + \lambda_3) \\
& \mu_{f22}(f_{22}(\boldsymbol{x})) = [f_{22}(\boldsymbol{x}) - f_{22}'] / [f_{22}^{L} - f_{22}'] \\
& \qquad = (-x_1 - y_1 + 3y_2 - y_3)/30 \geq (\lambda + \lambda_4) \\
& \mu_{f23}(f_{23}(\boldsymbol{x})) = [f_{23}(\boldsymbol{x}) - f_{23}'] / [f_{23}^{L} - f_{23}'] \\
& \qquad = (-x_1 - y_1 - y_2 + 3y_3)/30 \geq (\lambda + \lambda_5) \\
& \lambda + \lambda_i \leq 1, \; i = 1, 2, 3, 4, 5 \\
& \lambda \in [0, 1]
\end{aligned}$$

Solving the above compensatory model, we obtained a satisfactory solution: $\boldsymbol{f}^* = (f_{11}^*, f_{21}^*, f_{22}^*, f_{23}^*) = (31.07, 6.43, 6.43, 6.43)$ with satisfaction $\mu_{and} = 0.28$ at $\boldsymbol{x}^* = (1.07, 7.5, 7.5, 7.5)$ with $\gamma = 0.5$.

To investigate the influence of the compensatory operator, 11 cases were solved. The value of the parameter γ varies from 0 to 1. The results are listed in Table 7.3. From this table, we can see that the non-compensatory solution, or $\gamma = 1$, gives a value of $\lambda = 0.21$, which is very low. Thus, compensation improves the total degree of satisfaction.

Table 7.3. Results of Example 7.3

γ	μ_{and} λ	x_1, y_1, y_2, y_3	f_1, f_{21} f_{22} f_{23}
0.0	0.41 0.0	3.75,6.25,8.75,6.25	33.75, 0, 10, 0
0.1	0.37 0.0	3.75,6.25,8.75,6.25	33.75, 0, 10, 0
0.2	0.34 0.17	2.5, 7.5, 7.5, 7.5	32,5, 5, 5, 5
0.3	0.32 0.17	2.5, 7.5, 7.5, 7.5	32.5, 5, 5, 5
0.4	0.30 0.17	2.5, 7.5, 7.5, 7.5	32.5, 5, 5, 5
0.5	0.28 0.21	1.07,7.5, 7.5, 7.5	31.07,6.43,6.43,6.43
0.6	0.27 0.21	1.07,7.5, 7.5, 7.5	31.07,6.43,6.43,6.43
0.7	0.25 0.21	1.07,7.5, 7.5, 7.5	31.07,6.43,6.43,6.43
0.8	0.24 0.21	1.07,7.5, 7.5, 7.5	31.07,6.43,6.43,6.43
0.9	0.23 0.21	1.07,7.5, 7.5, 7.5	31,07,6.43,6.43,6.43
1.0	0.21 0.21	1.07,7.5, 7.5, 7.5	31.07,6.43, 6.43, 6.43

7.4 Bi-level Decentralized Problem with Unequally Important Objectives

Consider a bi-level problem with *s* decision units at the second level and assign the weight of each decision unit according to the degree of importance of the unit. Let these assigned weights be represented by w_{21}, w_{22}, ..., w_{2s} where the second subscript represents the decision unit in the second level and, in order to be unbiased, the summation of all the weights must be equal to one, or, $w_{21} + w_{22} + ... + w_{2s} = 1$. Using these weights in Equation 7.11, we obtain the following expression:

$$\max\ (\mu_{and})_2 = \lambda + (1-\gamma)[\lambda_1 + \lambda_2 + s\,(w_{21}\,\lambda_3 + w_{22}\,\lambda_4 + ... + w_{2s}\,\lambda_{s+2})\,]/(s+2) \tag{7.12}$$

subject to:

$$x \in Y$$
$$w_{21} + w_{22} + ... + w_{2s} = 1$$
$$0 \le w_{2\text{l}i} \le 1$$

where Y represents the original constraint set of Equation 7.11.

If we set the weights $w_{21} = w_{22} = ... = w_{2s} = 1/(s + 2)$, this general system reduces to Equation 7.11. One of the problems in assigning weight is that the assigned weights must be consistent. Various techniques, such as the simple additive weighting method, eigenvector method, weighted least square method, and entropy method, have been developed for assigning weight. For a detailed discussion, the reader can consult the various books on multiple attribute decision making [Hwang and Yoon, 1981]. Thus, after the importance of the objectives is ranked or ordered, we can use any of the above mentioned approaches to assign the weights.

Example 7.4. Consider the same problem solved in Example 7.3. Except, now, that the upper-level DM assigns different degrees of importance for the different divisions in the second level. The upper-level DM has decided that objective f_{21} is three times more important than either the second objective f_{22} or the third objective f_{23} and that the second objective is twice as important as the third objective. A matrix is usually used to represent these paired comparisons of importance and the relative weights can be obtained by using one of the methods listed above. In the present example, Saaty's eigenvector method [1990] will be adopted. The matrix based on assigned importance can be formed as:

$$A = [a_{ij}] = \begin{bmatrix} \frac{w_{21}}{w_{21}} & \frac{w_{21}}{w_{22}} & \frac{w_{21}}{w_{23}} \\ \frac{w_{22}}{w_{21}} & \frac{w_{22}}{w_{22}} & \frac{w_{22}}{w_{23}} \\ \frac{w_{23}}{w_{21}} & \frac{w_{23}}{w_{22}} & \frac{w_{23}}{w_{23}} \end{bmatrix} = \begin{bmatrix} \frac{1}{1} & \frac{3}{1} & \frac{3}{1} \\ \frac{1}{3} & \frac{1}{1} & \frac{2}{1} \\ \frac{1}{3} & \frac{1}{2} & \frac{1}{1} \end{bmatrix} \tag{7.13}$$

where a_{ij} represents the number of times greater is the importance of objective i than that of objective j.

The relative weights of the different objectives can be obtained by solving the matrix equation

$$(A - \lambda \mathbf{I})\, w = \mathbf{0} \tag{7.14}$$

where A is the 3 × 3 matrix defined in Equation 7.13 and λ is the eigenvalue of the matrix. The solution of Equation 7.14 results in three λ values. Substituting the maximum value of these three solutions into the matrix equation, we obtain three linear algebraic equations. The weights, w_{21}, w_{22}, and w_{23} can be obtained by solving this system of linear equations.

For the current problem, the weights obtained are: $(w_{21}, w_{22}, w_{23}) = (0.59, 0.25, 0.16)$. Substituting these weights into Equation 7.12, we have:

$$\max\ (\mu_{\text{and}})_2 = \lambda + (1-\gamma)\ [\lambda_1 + \lambda_2 + 3\ (0.59\ \lambda_3 + 0.25\ \lambda_4 + 0.16\ \lambda_5)]\ /\ 5$$

subject to:

$$\begin{aligned}
&\boldsymbol{x} \in \boldsymbol{X} \\
&\mu_{f11}(f_{11}(\boldsymbol{x})) = [f_{11}(\boldsymbol{x}) - f_{11}{}'] / [f_{11}{}^{L} - f_{11}{}'] \\
&\qquad = (x_1 + y_1 + 2\,y_2 + y_3) / 35 \geq \lambda + \lambda_1 \\
&[x_1 - (x_1{}^{U} - p_1)\,] / p_1 = x_1 / 5 \geq (\lambda + \lambda_2\,) \\
&x_1 \leq 10 \\
&\mu_{f21}(f_{21}(\boldsymbol{x})) = [f_{21}(\boldsymbol{x}) - f_{21}{}'] / [f_{21}{}^{L} - f_{21}{}'] \\
&\qquad = (-x_1 + 3\,y_1 - 2\,y_2 - y_3) / 30 \geq (\lambda + \lambda_3) \\
&\mu_{f22}(f_{22}(\boldsymbol{x})) = [f_{22}(\boldsymbol{x}) - f_{22}{}'] / [f_{22}{}^{L} - f_{22}{}'] \\
&\qquad = (-x_1 - y_1 + 3\,y_2 - y_3) / 30 \geq (\lambda + \lambda_4) \\
&\mu_{f23}(f_{23}(\boldsymbol{x})) = [f_{23}(\boldsymbol{x}) - f_{23}{}'] / [f_{23}{}^{L} - f_{23}{}'] \\
&\qquad = (-x_1 - y_1 - y_2 + 3\,y_3) / 30 \geq (\lambda + \lambda_5) \\
&\lambda + \lambda_i \leq 1,\ \ i = 1, 2, 3, 4, 5 \\
&w_{21} + w_{22} + \ldots + w_{2s} = 1 \\
&\lambda \in [0, 1]
\end{aligned}$$

where X represents the original constraint set, $f_{11}{}^{L}$ and $f_{11}{}'$ are the upper and lower bounds of f_{11}, and γ is the grade of compensation.

The satisfactory solution obtained is:

$f^* = (f_{11}^*, f_{21}^*, f_{22}^*, f_{23}^*) = (31.73, 8.85, 5.77, 5.77)$, $\mu_{\text{and}}^+ = 0.28$,
$x^* = (0.96, 8.27, 7.5, 7.5)$, $\gamma = 0.5$.

7.5 Multiple-level Decentralized Problem

The same procedure can be applied to multi-level problem with multiple followers. To simplify the discussion, let us consider a problem with three levels. Assume that there are s decision units in the second level and q decision units in the third level, then the objective function of the third-level auxiliary problem with equally important objectives can be illustrated as:

$$(\mu_{\text{and}})_3 = \lambda + (1-\gamma)\,[\lambda_1 + \lambda_2 + (\lambda_3 + \ldots + \lambda_{s+2}) + (\lambda_{s+3} + \ldots + \lambda_{2s+2}) + (\lambda_{2s+3} + \ldots + \lambda_{2s+q+2})]/(2s + q + 2) \tag{7.15}$$

where $(\lambda_3, \ldots, \lambda_{s+2})$ represent the goals of the second levels, $(\lambda_{s+3} + \ldots + \lambda_{2s+2})$ represent the decisions of the second level, and $(\lambda_{2s+3} + \ldots + \lambda_{2s+q+2})$ represent the goals of the third level.

For unequally important objectives, assuming the assigned weights are: w_{21}, w_{22}, ..., w_{2s} where $w_{21} + w_{22} + \ldots + w_{2s} = 1$, for the second level, and w_{31}, w_{32}, ..., w_{3q}, where $w_{31} + w_{32} + \ldots + w_{3q} = 1$, for the third level. The auxiliary model of the objective function for the third level can be represented as:

$$(\mu_{\text{and}})_4 = \lambda + (1-\gamma)[\lambda_1 + \lambda_2 + s(w_{21}\lambda_3 + \ldots + w_{2s}\lambda_{s+2})/2 + s(w_{21}\lambda_{s+3} + \ldots + w_{2s}\lambda_{2s+2})/2 + q(w_{31}\lambda_{2s+3} + \ldots + w_{3q}\lambda_{2s+q+2})] / (2s + q + 2) \tag{7.16}$$

where $(\lambda_3, \ldots, \lambda_{s+2})$ represent the goals of the second level, $(\lambda_{s+3} + \ldots + \lambda_{2s+2})$ represent the decisions of the second level, and $(\lambda_{2s+3} + \ldots + \lambda_{2s+q+2})$ represent the goals of the third level.

Obviously, the proposed approach can be extended easily to problem with k levels

7.6 Fuzzy Multi-level Problem

The multi-level problems considered in the previous chapters are really crisp problems and the problem itself is not fuzzy. The fuzziness comes from the mutual interactions between the various levels or between the divisions in the same level. In this section, we wish to consider fuzzy multi-level problems, where the coefficients are fuzzy or not well defined.

As discussed in Chapter 5, many different approaches have been proposed to solve programming problems with fuzzy coefficients. In principle, any of these approaches can be incorporated into the algorithms discussed for multi-level decision making in the previous chapters. However, because of space limitations, we shall only consider the approach discussed in Section 5.4 and combine it with the compensatory algorithm discussed early in this chapter. This is best illustrated by an example.

Example 7.5. Consider the problem solved in Example 5.5. Instead of two parallel objective functions, we shall consider these two objectives as in two different levels. Furthermore, we shall assume that DM in level one controls x_1 and DM in level two controls x_2. Thus, we have:

$$\max_{x1} f_1 = (1.5,2,2,2.5)\, x_1 + (-1.5,-1,-1,-0.5)\, x_2 \qquad \text{(upper level)}$$

where x_2 solves:

$$\max_{x2} f_2 = (0.8,1,1,1.4)\, x_1 + (1.5,2,1.5,2.5)\, x_2$$

(lower level)

subject to:

$$\begin{aligned}
&(2.5,3,3,3.5)\, x_1 + (-5.5,-5,-5,-4.5)\, x_2 \le (10, 25, 25, 26)\\
&(2.8,3,3.2,3.5)\, x_1 + (-1.5,-1,-1,-0.5)\, x_2 \le (15,20,25,35)\\
&(2.75,3,3.2,3.2)\, x_1 + (0.8,1,1,1.3)\, x_2 \le (20,25,25,42)\\
&(2.5,3,3,3.5)\, x_1 + (3.6,4,4,4.4)\, x_2 \le (30,32,35,50)\\
&(0.8,1,1,1.2)\, x_1 + (2.6,3,3,3.4)\, x_2 \le (14,15,16,25)\\
&x_1, x_2 \ge 0
\end{aligned}$$

The auxiliary problem formulated in Example 5.5 is reproduced in the following:

$$\min \{\theta^1{}_2, \theta^2{}_2, \delta_1, \delta_2, \delta_3, \delta_4, \delta_5\}$$

subject to:

$$\begin{aligned}
&\theta_{12} = (2.5x_1 - 0.5x_2 - f_1) / (0.5x_1 + 0.5x_2) \ge \alpha\\
&2x_1 - x_2 \le f_1 \le 2.5x_1 - 0.5x_2\\
&\theta_{22} = (1.4x_1 + 2.5x_2 - f_2) / (0.4 - x_1) \ge \alpha\\
&x_1 + 2.5x_2 \le f_2 \le 1.4x_1 + 2.5x_2\\
&\delta_1 = (26 - 2.5x_1 + 5.5x_2) / (11 + 0.5x_1 + 0.5x_2) \ge \alpha\\
&15 \le 3x_1 - 5x_2\\
&26 \ge 2.5x_1 - 5.5x_2\\
&\delta_2 = (35 - 2.8x_1 + 1.5x_2) / (10 + 0.2x_1 + 0.5x_2) \ge \alpha\\
&25 \le 3x_1 - x_2\\
&35 \ge 2.8x_1 - 1.5x_2\\
&\delta_3 = (42 - 2.75x_1 - 0.8x_2) / (17 + 0.25x_1 + 0.2x_2) \ge \alpha\\
&25 \le 3x_1 + x_2\\
&42 \ge 2.75x_1 + 0.8x_2\\
&\delta_4 = (50 - 2.5x_1 - 3.6x_2) / (15 + 0.5x_1 + 0.4x_2) \ge \alpha
\end{aligned}$$

$$35 \le 3x_1 + 4x_2$$
$$50 \ge 2.5x_1 + 3.6x_2$$
$$\delta_5 = (25 - 0.8x_1 - 2.6x_2) / (9 + 0.2x_1 + 0.4x_2) \ge \alpha$$
$$16 \le x_1 + 3x_2$$
$$25 \ge 0.8x_1 + 2.6x_2$$
$$\theta_{12}, \theta_{22}, \delta_1, \delta_2, \delta_3, \delta_4, \delta_5 \in [0, 1]$$
$$x_1, x_2 \ge 0$$

In order to solve the above two-level problem, let us assume that the fuzzy ranges for the two objectives are:

$$f_1 \in [0, 23.47],\ f_2 \in [0, 21.12]$$

We shall further assume that the first-level decision x_1 is around 11.6 with maximum uncertainty on the plus side of 2.06 and maximum uncertainty on the minus side of 1.94. Now, following the compensatory approach, the auxiliary problem can be obtained as:

$$\max \mu_{\text{and}} = \lambda + (1-\gamma)\ (\lambda_1+\lambda_2+\lambda_3) / 3$$

subject to:

$$\boldsymbol{x} \in \boldsymbol{Z}$$
$$(f_1 - 0) / (23.47 - 0) \ge \lambda + \lambda_1$$
$$(x_1 - 9) / (11.6 - 9) \ge \lambda + \lambda_2$$
$$(13 - x_1) / (13 - 11.6) \ge \lambda + \lambda_2$$
$$(f_2 - 0) / (21.12 - 0) \ge \lambda + \lambda_3$$
$$\gamma, \lambda, \lambda_1, \lambda_2\ \lambda_3 \in [0, 1]$$

where $\boldsymbol{x} \in \boldsymbol{Z}$ represents the set of original constraint, and γ is the degree of compensation.

This problem can be solved easily and the solution is:

$$\boldsymbol{f}^* = (f_1^*, f_2^*) = (22.62, 19.24)$$
$$\boldsymbol{x}^* = (x_1^*, x_2^*) = (10.88, 2.48)$$
$$\mu_{\text{and}} = 0.920, \gamma = 0.5$$
$$\alpha = 0.5$$

7.7 Discussions

The mutually interactive decision concept is a powerful approach. However, only a beginning has been accomplished. Many more detailed investigations are needed. Two of the most crucial issues are how to compromise or aggregate between the different DMs and how to represent the information of the different levels if this information is vague, not well defined, or even conflicting. The first problem is handled by the use of different aggregation operators and the second one used the possibility concept. Only the Werners aggregation operator is used. Obviously, many other existing aggregation operators should also be investigated.

Even within the Werners operator, further investigation is needed. For example, the value of γ, which controls the size of the compensatory part, needs more detailed study.

In order to reduce the problem to linear form, we have assumed that the DM can furnish the desired possibility level. If the possibility is assumed unknown, then we must solve a non-linear problem. One way to solve this non-linear problem is by the use of exhaustive search for all the possibility levels. In other words, first, assume all possible values of possibility and, then, form different linear problems with each assumed possibility value. In this way, we are essentially solving a series of linear problems.

Chapter 8

Possibilistic Minimum-cost Flow Problem

To illustrate the applications of the proposed fuzzy approaches, the minimum-cost flow (MCF) problem was investigated and several examples were solved. Minimum-cost flow problem is a general form of network flow problem, whose aim is to find the least cost of shipments of commodities through capacitated network in order to satisfy demands at certain nodes from available supplies at other nodes. The results from the study of this general problem can be applied to many other network problems such as transportation, maximum flow, assignment, shortest path, and transshipment problems. In addition, the results from these kind of study frequently offer bounds or rough estimates to more complicated network problems such as the Chinese postman problem [Pearn 1994]. Therefore, the MCF problem plays a crucial role in the understanding of network flow. Furthermore, the MCF is also very practical and has been applied to a number of real-world problems such as multi-stage production and inventory planning, mold allocation, nurse scheduling, project assignment, and automobile routing [Ahuja *et al.* 1993].

One problem with the application of the network algorithm is the fact that the actual encountered problem is frequently vague, uncertain, not well defined, or even conflicting. Thus, the fuzzy concept appears to be an ideal tool to help us to overcome these difficulties. It should be emphasized that the fuzzy approach can solve the original problem *as it is* without any simplification or modification. Furthermore, the original problem represented by the fuzzy or possibilistic concept is frequently in much simpler in form and thus can be solved much more easier.

As discussed in Chapter 5, there are many different formulations and approaches for solving the fuzzy programming problem. We shall follow the possibility approach because of its practical value and can represent linguistic expressions in programming form with constraints. To reduce the complexity of the problem, we shall use possibility programming to handle the vagueness of the parameters and the *max-min* aggregation as was original proposed by Bellman and Zadeh [1970] to handle the multi-level aspects. To overcome the deficiencies of the *max-min* approach, which is non-compensatory, a compensatory operator due to Werners [1988] is used for aggregation in the last section.

Negi and Lee [1989, 1993] proposed approaches to overcome the problem of resulting non-linear equations by assuming the desired possibility is given *a priori*. This assumption not only simplifies the original problem but also is more realistic. Depending on the degree of importance of the problem, the DM should be able to specify the desired possibility level. For example, if the problem is important, then we would like to have a higher reliability, which can be obtained by requiring a higher possibility value. These authors used the more general trapezoidal fuzzy number to formulate the possibility linear programming. However, the formulation will generate too many constraints, which enlarge the problem considerably. To overcome this problem, the fuzzy range or limit is specified. In this way, we can eliminate at least one-third of the capacity constraints.

In actual practice, the DM is usually required to consider many different objectives. Thus, we shall formulate and solve a multi-objective MCF network. We also wish to consider a new decision-making environment, where multiple DMs in a hierarchy make the mutually satisfactory decisions. In other words, we extend the supervised search procedure for the multi-level system discussed in the early chapters to solve the fuzzy multi-level MCF problem. This proposed approach offers a new direction for solving the various network problems such as the network design problem and transportation planning.

In the following sections, fuzzy MCF problem is first defined. Then, fuzzy representation of capacity constraints and the possibility approach for the MCF problem are formulated. Finally, multi-objective and multi-level decision environments are investigated. Several numerical examples are solved to illustrate the approach.

8.1 Minimum-cost Flow Problem

Minimum-cost flow problem is to find the minimum cost of flow from the supply nodes to the demand notes through capacitated and directed arcs. Various researchers have investigated this general problem. In this short introduction, only the basic notations and equations will be summarized. For a detailed discussion, the reader can consult the literature [Ahuja *et al.* 1993, Murty 1992].

To establish the equations for the MCF problem, let us consider a directed network, $G\,(N, A)$, which is composed of directed arcs, A, and nodes, N. Each arc, $(i, j) \in A$, has a capacity, which has an upper limit or bound capacity u_{ij} and lower bound ℓ_{ij}. Each arc is also associated with a cost c_{ij}. Each node, $i \in N$, possesses a number of resources, $b\,(i)$. The node can be separated into supply node, demand node, or transient node depending on whether $b(i) > 0$, $b(i) < 0$, or $b(i) = 0$, respectively. The crisp MCF problem can now be formulated as:

$$\min f(x) = \sum_{(i,j)\in A} c_{ij} x_{ij} \tag{8.1}$$

subject to:

$$\sum_{j:(i,j)\in A} x_{ij} - \sum_{j:(j,i)\in A} x_{ji} = b(i), \forall i \in N$$
$$\ell_{ij} \le x_{ij} \le u_{ij}, \forall (i,j) \in A$$
$$x_{ij} \ge 0, \forall (i,j) \in A$$

where $i = 1, 2, ..., m, j = 1, 2, ..., n$, and all x_{ij} are integers.

The objective is to minimize the total cost of flow. The first constraint in Equation 8.1 is the constraint for the conservation of flow. It maintains balance between the input and the output at each node. The second constraint in Equation 8.1 is the capacity constraint associated with the lower and the upper bounds of each arc. And the third constraint ensures non-negative flow at each arc. In general, the MCF problem also has several additional assumptions such as: (a) all data must be integers; (b) the network is directed; (c) the supply/demand at each node satisfies the condition $\sum b(i) = 0$; (d) all arc costs are non-negative; and (e) the MCF has a feasible solution. Although we can alleviate some assumptions later, these assumptions guarantee the existence and optimality of the solution.

Equation 8.1 represents a fairly general network problem. For example, the above equation reduces to a maximum flow problem if the second constraint is removed, all the flow costs obey $c_{ij} = 1$, and the problem is to maximize the objective function. Another important network problem is the transportation problem, the above equation reduces to this problem if the network is bipartite and all arcs are directed from a source to a sink without capacity restriction

Investigations in fuzzy networks are fairly limited. Chanas and Kolodziejczyk [1982, 1984, 1986] studied the maximum flow problem. More recently, Chanas and coworkers [1995] investigated fuzzy optimal flow in imprecise structures and the transportation network. ÓhÉigeartaigh [1982] developed a transportation algorithm with fuzzy capacities based on Zimmermann's approach [1978]. Both Chanas *et al.* [1984] and Delgado *et al.* [1987] proposed parametric approach to the transportation problem. Recently, Bit *et al.* [1992, 1993a, 1993b] made a series of studies on fuzzy transportation. Ignizio and Daniels [1983] studied the fuzzy generalized network and fuzzy multi-criteria programming and Klein [1991] developed an algorithm for fuzzy shortest path problems based on dynamic programming

8.2 Possibility Approach to Minimum-cost Flow Problem

To establish the fuzzy MCF model, we shall assume that both the capacity and the cost are fuzzy. Thus, we have:

$$\min \tilde{f}(x) = \sum_{(i,j)\in A} \tilde{c}_{ij} x_{ij} \tag{8.2}$$

subject to:

$$\sum_{j:(i,j)\in A} x_{ij} - \sum_{j:(j,i)\in A} x_{ji} = b(i), \forall i \in N$$
$$\tilde{\ell}_{ij} \le x_{ij} \le \tilde{u}_{ij}, \forall (i,j) \in A$$
$$x_{ij} \ge 0, \forall (i,j) \in A$$

where $\tilde{f}(x)$ represents the fuzzy objective, $\tilde{c}_{ij}$ represents the fuzzy cost, and $\tilde{l}_{ij}$ and $\tilde{u}_{ij}$ represent the fuzzy lower and fuzzy upper bounds of each arc, respectively, $i = 1, 2, ..., m, j = 1, 2, ..., n,$ and all x_{ij} are integers. Trapezoidal fuzzy number, which is in the form of (n_1, n_2, n_3, n_4), will be assumed in the following. The membership function is one within $[n_2,n_3]$ and reduces to zero at n_1 and n_2.

8.2.1 Capacity Constraint Modification

Negi and Lee [1993] formulated the possibility programming based on the trapezoidal fuzzy numbers and by the use possibility ranking of fuzzy numbers due to Dubois and Prade [1988]. There are two problems in applying this formulation to the MCF network. The first is the increase in the number of constraints and the second is the difficulty in defining the fuzzy membership function for the capacity constraint set. To avoid these difficulties, we adopted the concept of fuzzy range instead of the possibilistic range in formulating the arc capacity, where the value of the fuzzy range is controlled by a cut-off parameter, α.

Using the possibility formulation of Negi and Lee, the original m constraints will increase to $(3m + 3)$ constraints. This large number of constraint increase will make the problem difficult to solve. This is especially true for network problems, where fairly large sparse matrix is generally needed to represent even a reasonably large network.

The second difficulty can be explained easily by the use of an example. For trapezoidal fuzzy numbers, the arc capacity constraint equation, or the second constraint in Equation 8.2 can be expressed as:

$$(1, 1, 1, 1)\, x_{ij} \le \tilde{b}_i = (b_{i1}, b_{i2}, b_{i3}, b_{i4}) = (4, 4, 5, 7.5) \tag{8.3}$$

We only used the second inequality in the second constraint of (8.2) and assumed $\tilde{b}_i$ as the upper limit, which is assumed to have a value of (4, 4, 5, 7.5). As discussed in Chapter 5, if we use the exceedence possibility, Equation 5.40, the value of the membership function can be calculated as:

$$\begin{aligned}\delta &= (b_4 - r_1) / [(r_2 - r_1) + (b_4 - b_3)] \\ &= (7.5 - 1) / [(4 - 4) + (1 - 1)] = 6.5 / 0 = \infty\end{aligned}$$

Thus, this membership function is not defined. In fact, we can show that for any left-right membership function, the above situation would happen if one side of the fuzzy number is crisp, or the membership function drop suddenly from one to zero.

To avoid this difficulty, we shall formulate the capacity constraint directly from a cut-off parameter, α, based on the particular characteristics of the membership function.

For the membership function in the MCF network, there exists a lower bound l_{ij} and an upper bound u_{ij} for each arc capacity. Thus, for a membership function of the form, $\tilde{b}_i = (l_{ij1}, l_{ij2}, u_{ij1}, u_{ij2})$, we can define the membership function as:

$$\mu(x_{ij}) = \begin{cases} \delta_1 = \dfrac{x_{ij} - \ell_{ij1}}{\ell_{ij2} - \ell_{ij1}}, \text{ if } \ell_{ij1} \le x_{ij} \le \ell_{ij2} \\ 1, \text{ if } \ell_{ij2} \le x_{ij} \le u_{ij1} \\ \delta_2 = \dfrac{u_{ij2} - x_{ij}}{u_{ij2} - u_{ij1}}, \text{ if } u_{ij1} \le x_{ij} \le u_{ij2} \\ 0, \text{ otherwise} \end{cases} \tag{8.4}$$

where the left-hand-side of the interval, $\tilde{l}_{ij}$, is represented by $[l_{ij1}, l_{ij2}]$; and the right-hand-side of the interval, $\tilde{u}_{ij}$, is represented by $[u_{ij1}, u_{ij2}]$ with δ_1 and $\delta_2 \in [0, 1]$.

Now, let us define a cut-off parameter α, whose value is restricted to the range [0, 1], or $0 \le \alpha \le 1$. Since we wish to seek the maximum possibility of $\delta_i \ge \alpha$, $i = 1, 2$, based on the cut-off parameter, the left-hand-side limit or bound can be represented as:

$$\delta_1 = (x_{ij} - l_{ij1}) / (l_{ij2} - l_{ij1}) \ge \alpha$$

after multiply out, we have:

$$x_{ij} \ge \alpha\, (l_{ij2} - l_{ij1}) + l_{ij1}$$

in the same way, we can obtain the right-hand-side limit as:

$$\delta_2 = (u_{ij2} - x_{ij}) / (u_{ij2} - u_{ij1}) \ge \alpha$$

or

$$x_{ij} \le u_{ij2} - \alpha\, (u_{ij2} - u_{ij1})$$

Combining both sides, we finally have the desired possibility fuzzy interval:

$$\alpha\, (l_{ij2} - l_{ij1}) + l_{ij1} \le x_{ij} \le u_{ij2} - \alpha\, (u_{ij2} - u_{ij1}) \tag{8.5}$$

The cut-off parameter α controls the feasibility or possibility of the arc capacity. Increase the cut-off parameter increases the value of the membership function or the possibility. At the same time, this increase reduces the interval of flow. In other words, since the possibility or the aspiration level is increased, we must pay a higher price for it. The effect of the cut-off parameter is somewhat similar to the parametric approach proposed by Carlsson and Korhonen [1986]. To illustrate the approach in more detail, consider the following arc capacity:

$$\widetilde{b}_i = [l_{ij1}, l_{ij2}, u_{ij1}, u_{ij2}] = (0, 0, 2, 3)$$

and assuming the cut-off parameter $\alpha = 0.5$, then the possibility distribution $\delta_2 \geq \alpha$ for minimum satisfaction will be

$$\delta_2 = (u_{ij2} - x_{ij}) / (u_{ij2} - u_{ij1}) \geq \alpha = 0.5$$

Simplify and use the given numerical values, we obtain $x_{ij} \leq 2.5$. In this case, the fuzzy boundary varies within the interval [2, 3] depending on the value of α. If a higher possibility is desired, The vagueness will be smaller or the interval will be narrower. Thus, the value of the cut-off parameter controls the tolerance we wish to allow.

From computational standpoint, the major advantage of using Equation 8.5 is to reduce the number of constraints. The approach reduces one-third of the number of capacity constraints required. Furthermore, if the lower bound of the arc l_{ij} is zero, we can reduce another one-third, because the system constraint, $x_{ij} \geq 0$, would replace the lower bound zero constraint.

8.2.2 Possibility Programming

Using the same method used to obtain Equation 5.51, except replacing the possibility ranking, Equation 5.41, with the cut-off parameter Equation 8.5, the following possibility programming problem can be obtained for the fuzzy MCF problem represented by Equation 8.2:

$$\min f(x) \tag{8.6}$$

subject to:

$$\begin{aligned} &\theta_1 \geq \alpha \\ &\alpha\,(l_{ij2} - l_{ij1}) + l_{ij1} \leq x_{ij} \leq u_{ij4} - \alpha\,(u_{ij2} - u_{ij1}), \ \forall\, i \text{ and } j \\ &\Sigma_i \Sigma_j\, c_{ij1}\, x_{ij} \leq f \leq \Sigma_i \Sigma_j\, c_{ij2}\, x_{ij} \\ &\theta_1, \alpha \in [0, 1] \\ &x_{ij} \geq 0 \end{aligned}$$

where $i = 1, 2, ..., m, j = 1, 2, ..., n,$ and all x_{ij} are integers. Notice that the above equation only has 2m constraints, instead of 3m as in Equation 5.51.

This integer programming problem can be solved by an existing mixed-integer code. In case there is no feasible solution, adjustments of the cut-off parameter will be needed. Thus, an interactive procedure can be adopted to adjust the range of the compromise.

Example 8.1. A fuzzy MCF network with 8 nodes and 11 arcs was solved in this example. The structure of the network and the corresponding nodes and arcs are listed in Table 8.1, where the heading S/D represents "supply/demand". The other headings are self explanatory. Let x_{ij} represents the flow from node i to node j. The directional flows x_{21}, x_{23}, x_{26}, x_{14}, x_{34}, x_{35}, x_{47}, x_{56}, x_{57}, x_{68}, and x_{78} flow through the arcs ①, ②, ③, ④, ⑤, ⑥, ⑦, ⑧, ⑨, ⑩, and, ⑪ respectively.

Table 8.1. Parameters for Example 8.1

node	S/D, arc	Fuzzy cost	Fuzzy capacity	Note
1	10, 1	0.5, 1, 1.5, 2	0, 0, 9, 11	x_{21}
2	20 2	0, 0, 0.5, 1	0, 1, 9, 13	x_{23}
3	0, 3	5, 6, 7, 8	0, 0, 9, 11	x_{26}
4	-5, 4	1.5, 2, 2.5, 3	0, 2, 15, 16	x_{14}
5	0, 5	0.5, 1, 1.5, 2	0, 0, 5, 9.5	x_{34}
6	0, 6	3, 4, 5, 6	0, 0, 10, 12	x_{35}
7	-15, 7	4, 5, 6, 7	0, 2, 10, 14.5	x_{47}
8	-10, 8	1.5, 2, 2.5, 3.5	0, 0, 20, 22	x_{56}
	9	6, 7, 8, 9	0, 0, 15, 17	x_{57}
	10	7, 8, 9. 10	0, 1, 10, 12	x_{68}

The MCF problem can be formulated as:

$$\min f1 = (0.5, 1, 1.5, 2)\, x_{21} + (0, 0, 0.5, 1)\, x_{23} + (5, 6, 7, 8)\, x_{26} + (1.5, 2, 2.5, 3)\, x_{14} + (0.5, 1, 1.5, 2)\, x_{34} + (3, 4, 5, 6)\, x_{35} + (4, 5, 6, 7)\, x_{47} + (1.5, 2, 2.5, 3.5)\, x_{56} + (6, 7, 8, 9)\, x_{57} + (7, 8, 9, 10)\, x_{68} + (8, 9, 10, 11.5)\, x_{78}$$

subject to:

$$x_{14} + x_{21} = 10$$
$$x_{21} + x_{23} + x_{26} = 20$$
$$x_{34} + x_{35} - x_{23} = 0$$
$$x_{47} - x_{14} - x_{34} = -5$$
$$x_{56} + x_{57} - x_{35} = 0$$
$$x_{68} - x_{56} - x_{26} = 0$$
$$x_{78} - x_{47} - x_{57} = -15$$
$$- x_{68} - x_{78} = -10$$
$$x_{21} \in (0, 0, 9, 11)$$
$$x_{23} \in (0, 1, 9, 13)$$
$$x_{26} \in (0, 0, 9, 11)$$
$$x_{14} \in (0, 2, 15, 16)$$
$$x_{34} \in (0, 0, 5, 9.5)$$
$$x_{35} \in (0, 0, 10, 12)$$
$$x_{47} \in (0, 2, 10, 14.5)$$
$$x_{56} \in (0, 0, 20, 22)$$
$$x_{57} \in (0, 0, 15, 17)$$
$$x_{68} \in (0, 1, 10, 12)$$

$$x_{78} \in (0, 0, 15, 16.5)$$

where all the x_{ij} are integers.

Using Equation 8.6, the auxiliary problem can be obtained as:

$$\min fl$$

subject to:

$$(fl - 0.5\,x_{21} - 5\,x_{26} - 1.5\,x_{14} - 0.5\,x_{34} - 3\,x_{35} - 4\,x_{47} - 1.5\,x_{56} - 6\,x_{57} - 7\,x_{68} - 8\,x_{78}) \,/\, (0.5\,x_{21} + x_{26} + 0.5\,x_{14} + 0.5\,x_{34} + x_{35} + x_{47} + x_{68} + x_{78}) \geq \alpha$$
$$0.5\,x_{21} + 5\,x_{26} + 1.5\,x_{14} + 0.5\,x_{34} + 3\,x_{35} + 4\,x_{47} + 1.5\,x_{56} + 6\,x_{57} + 7\,x_{68} + 8\,x_{78} \leq fl$$
$$fl \leq x_{21} + 6\,x_{26} + 2\,x_{14} + x_{34} + 4\,x_{35} + 5\,x_{47} + 2\,x_{56} + 7\,x_{57} + 8\,x_{68} + 9\,x_{78}$$
$$x_{14} + x_{21} = 10$$
$$x_{21} + x_{23} + x_{26} = 20$$
$$x_{34} + x_{35} - x_{23} = 0$$
$$x_{47} - x_{14} - x_{34} = -5$$
$$x_{56} + x_{57} - x_{35} = 0$$
$$x_{68} - x_{56} - x_{26} = 0$$
$$x_{78} - x_{47} - x_{57} = -15$$
$$- x_{68} - x_{78} = -10$$
$$x_{21} \leq 11 - 2\alpha$$
$$\alpha \leq x_{23} \leq 13 - 4\alpha$$
$$x_{26} \leq 11 - 2\alpha$$
$$2\alpha \leq x_{14} \leq 16 - \alpha$$
$$x_{34} \leq 9.5 - 4.5\alpha$$
$$x_{35} \leq 12 - 2\alpha$$
$$2\alpha \leq x_{47} \leq 14.5 - 4.5\alpha$$
$$x_{56} \leq 22 - 2\alpha$$
$$x_{57} \leq 17 - 2\alpha$$
$$\alpha \leq x_{68} \leq 12 - 2\alpha$$
$$x_{78} \leq 16.5 - 1.5\alpha$$
$$\alpha \in [0, 1]$$

where all x_{ij} are positive integers.

This transformed problem was solved by using the LINGO mixed-integer code [1992] with a cut-off parameter $\alpha = 0.5$. The optimal solution obtained is:

$$fl = 236.5,\ x_{14} = 10,\ x_{34} = 7,\ x_{35} = 4,\ x_{47} = 12,\ x_{57} = 3,\ x_{68} = 10,\ x_{26} = 9,\ x_{23} = 11,\ x_{56} = 1,\ x_{78} = x_{21} = 0.$$

8.3 Possibility Approach to Multi-objective Minimum-cost Flow Problem

We can extend the formulation of Equation 8.1 into the following multi-objective problem:

$$\min/\max\ f^k(x) = \sum_{(i,j)\in A} c^k{}_{ij} x_{ij},\ k = 1,2,\ldots,K \tag{8.7}$$

subject to:

$$\sum_{j:(i,j)\in A} x_{ij} - \sum_{j:(j,i)\in A} x_{ji} = b(i),\ \forall i \in N$$
$$\ell_{ij} \le x_{ij} \le u_{ij},\ \forall (i,j) \in A$$
$$x_{ij} \ge 0,\ \forall (i,j) \in A$$

where $i = 1, 2, \ldots, m, j = 1, 2, \ldots, n,$ and all x_{ij} are integers.

We shall use the compromise programming approach with ideal and anti-ideal solutions as that discussed in Section 5.2.1 to solve this multi-objective problem. Following the discussions in that section, the above problem is transformed into:

$$\min \left\{ \sum_k \left[\frac{f^k(x^*) - f^k(x)}{f^k(x^*) - f^k(x^-)} \right]^p \right\}^{\frac{1}{p}} \tag{8.8}$$

subject to:

$$\sum_{j:(i,j)\in A} x_{ij} - \sum_{j:(j,i)\in A} x_{ji} = b(i),\ \forall i \in N$$
$$\ell_{ij} \le x_{ij} \le u_{ij},\ \forall (i,j) \in A$$
$$x_{ij} \ge 0,\ \forall (i,j) \in A$$

where $i = 1, \ldots, m, j = 1, \ldots, n.$ $f^k(x^*)$ and $f^k(x^-)$ are the ideal and anti-ideal solutions of the kth objective function. Substituting the constraints in Equation 8.6 into Equation 8.8, we obtain the fuzzy multi-objective MCF problem with distance function $h(x)$:

$$\min h(x) = \{\Sigma_k [\, (f^k(x^+) - f^k(x)\,) \,/\, (f^k(x^+) - f^k(x^-)\,)\,]^p \}^{1/} \tag{8.9}$$

subject to:

$$\alpha\,(l_{ij2} - l_{ij1}) + l_{ij1} \le x_{ij} \le u_{ij4} - \alpha\,(u_{ij2} - u_{ij1}),\ \forall\ i \text{ and } j$$
$$\Sigma_i \Sigma_j c_{ij1} x_{ij} \le f^1 \le \Sigma_i \Sigma_j c_{ij2} x_{ij}, \text{ and } \theta^1{}_1 \ge \alpha$$
$$\Sigma_i \Sigma_j c_{ij3} x_{ij} \le f^1 \le \Sigma_i \Sigma_j c_{ij4} x_{ij}, \text{ and } \theta^1{}_2 \ge \alpha$$
$$\ldots$$

$$\sum_i \sum_j c_{ij1}\, x_{ij} \le f^K \le \sum_i \sum_j c_{ij2}\, x_{ij}, \text{ and } \theta^K{}_1 \ge \alpha$$
$$\sum_I \sum_j c_{ij3}\, x_{ij} \le f^K \le \sum_i \sum_j c_{ij4}\, x_{ij}, \text{ and } \theta^K{}_2 \ge \alpha$$
$$x_{ij} \ge 0$$

where $i = 1, \ldots, m, j = 1, \ldots, n, k = 1, \ldots, K$, and all x_{ij} are integers.

Since α is, in general unknown, the above problem is a non-linear problem. As has been discussed in Chapter 5, if the value of α is given, the above problem becomes linear. In order to solve this problem, whether it is in linear or non-linear form, we must first obtain the ideal and anti-ideal solutions from the original equation. The following example illustrates the procedure involved.

Example 8.2. Consider the same network solved in Example 8.1, except there is an additional second objective, which minimizes the total passing time. There are 8 nodes and 11 arcs in the network and the data for the problem is summarized in Table 8.2, where the heading "S/D" represents "supply/demand.

Table 8.2. Parameters for Examples 8.2 and 8.3

Node	S/D	Arc	cost objective	Time object
1	10	1	0.5, 1, 1.5, 2	1, 2, 2.5, 3
2	20	2	0, 0, 0.5, 1	1, 2, 2.5, 3
3	0	3	5, 6, 7, 8	5, 6 7, 8
4	-5	4	1.5, 2, 2.5, 3	2, 2, 3, 3
5	0	5	0.5, 1, 1.5, 2	1.2, 2, 2, 2.5
6	0	6	3, 4, 5, 6	1.5, 2, 2, 2.5
7	-15	7	4, 5, 6, 7	6, 7, 7.5, 8
8	-10	8	1.5, 2, 2.5, 3.5	1, 2, 2.5, 3
		9	6, 7, 8, 9	1, 2, 2.5, 3
		10	7, 8, 9, 10	2, 2.5, 3, 3.5
		11	8, 9, 10, 11.5	2, 2.5, 3, 3.5

In order to obtain the ideal and anti-ideal solutions, we need to solve four single objective optimization problems. Let $f1$ and $f2$ represent the ideal solutions of the first, or the original objective, and the second, or the newly added objective, respectively, and $f3$ and $f4$ represent the anti-ideal solutions of the corresponding objectives. The ideal solution for the first objective was obtained in Example 8.1. We still need to obtain the other three solutions by solving the following three optimization problems.

$$\min f2 \tag{8.10}$$

subject to:

$(f2 - 2\,x_{21} - x_{23} - 5\,x_{26} - 2\,x_{14} - 1.2\,x_{34} - 1.5\,x_{35} - 6\,x_{47} - x_{56} - x_{57}$
$- 2\,x_{68} - 2\,x_{78}) \,/\, (x_{23} + x_{26} + 0.8\,x_{34} + 0.8\,x_{35} + 0.5\,x_{35} + x_{47} + x_{56} +$
$0.5\,x_{68} + 0.2\,x_{78}) \geq \alpha$
$2\,x_{21} + x_{23} + 5\,x_{26} + 2\,x_{14} + 1.2\,x_{34} + 1.5\,x_{35} + 6\,x_{47} + x_{56} + x_{57} + 2$
$x_{68} + 2\,x_{78} \leq f2$
$f2 \leq 2\,x_{21} + 2\,x_{23} + 6\,x_{26} + 2\,x_{14} + 2\,x_{34} + 2\,x_{35} + 7\,x_{47} + 2\,x_{56} + 2$
$x_{57} + 2.5\,x_{68} + 2.2\,x_{78}$,
$x_{14} - x_{21} = 10$
$x_{21} + x_{23} + x_{26} = 20$
$x_{34} + x_{35} - x_{23} = 0$
$x_{47} - x_{14} - x_{34} = -5$
$x_{56} + x_{57} - x_{35} = 0$
$x_{68} - x_{56} - x_{26} = 0$
$x_{78} - x_{47} - x_{57} = -15$
$- x_{68} - x_{78} = -10$
$x_{21} \leq 11 - 2\,\alpha$
$\alpha \leq x_{23} \leq 13 - 4\,\alpha$
$x_{26} \leq 11 - 2\,\alpha$
$2\,\alpha \leq x_{14} \leq 16 - \alpha$
$x_{34} \leq 9.5 - 4.5\,\alpha$
$x_{35} \leq 12 - 2\,\alpha$
$2\,\alpha \leq x_{47} \leq 14.5 - 4.5\,\alpha$
$x_{56} \leq 22 - 2\,\alpha$
$x_{57} \leq 17 - 2\,\alpha$
$\alpha \leq x_{68} \leq 12 - 2\,\alpha$
$x_{78} \leq 16.5 - 1.5\,\alpha$
$\alpha \in [0, 1]$

where all x_{ij} are integers. For simplicity in notation, the last 20 constraints, which include the 8 equality and the last 11 inequality constraints, plus the last $\alpha \in [0, 1]$ constraint, will be represented by $x \in X$.

$$\max f3 \qquad (8.11)$$

subject to:

$(2\,x_{21} + x_{23} + 8\,x_{26} + 3\,x_{14} + 2\,x_{34} + 6\,x_{35} + 7\,x_{47} + 3.5\,x_{56} + 9\,x_{57} +$
$10\,x_{68} + 11.5\,x_{78} - f3) \,/\, (0.5\,x_{21} + 0.5\,x_{23} + x_{26} + 0.5\,x_{14} + 0.5\,x_{34}$
$+ x_{35} + x_{47} + x_{56} + x_{57} + x_{68} + 1.5\,x_{78}) \geq \alpha$
$1.5\,x_{21} + 0.5\,x_{23} + 7\,x_{26} + 2.5\,x_{14} + 1.5\,x_{34} + 5\,x_{35} + 6\,x_{47} + 2.5\,x_{56}$
$+ 8\,x_{57} + 9\,x_{68} + 10\,x_{78} \leq f3$
$f3 \leq 2\,x_{21} + x_{23} + 8\,x_{26} + 3\,x_{14} + 2\,x_{34} + 6\,x_{35} + 7\,x_{47} + 3.5\,x_{56} + 9$
$x_{57} + 10\,x_{68} + 11.5\,x_{78}$
$x \in X$.

where x ∈ X. represents the 20 constraints discussed above.

$$\max f4 \qquad (8.12)$$

subject to:

$$(f4 - 3\,x_{21} - 3\,x_{23} - 8\,x_{26} - 3\,x_{14} - 3\,x_{34} - 2.5\,x_{35} - 8\,x_{47} - 3\,x_{57} - 3.5\,x_{68} - 4\,x_{78}) \,/\, (0.5\,x_{21} + 0.5\,x_{23} + x_{26} + 0.5\,x_{14} + 3\,x_{34} + 0.5\,x_{35} + 0.5\,x_{47} + 0.5\,x_{57} + 0.5\,x_{68} + x_{78}) \geq \alpha$$
$$2.5\,x_{21} + 2.5\,x_{23} + 7\,x_{26} + 2.5\,x_{14} + 3\,x_{34} + 2\,x_{35} + 7.5\,x_{47} + 2.5\,x_{56} + 2.5\,x_{57} + 3\,x_{68} + 3\,x_{78} \leq f4$$
$$f4 \leq 3\,x_{21} + 3\,x_{23} + 8\,x_{26} + 3\,x_{14} + 3\,x_{34} + 2.5\,x_{35} + 8\,x_{47} + 3\,x_{56} + 3\,x_{57} + 3.5\,x_{68} + 4\,x_{78},$$
$$x \in X$$

Using a cut-off parameter value, α = 0.5, the ideal solutions obtained are $f1$ = 236.50, $f2$ = 176.60, and the anti-ideal solutions are $f3$ = 409.75, $f4$ = 293.25. These solutions are listed in Table 8.3.

Table 8.3. Ideal and anti-ideal solutions for Example 8.2

	min f1	min f2	max f3	max f4	
Objective / Arc	236.50	176.60	409.75	293.25	Note
1	0	0	5	5	x_{21}
2	11	11	11	11	x_{23}
3	9	9	4	4	x_{26}
4	10	10	15	15	x_{14}
5	7	0	0	2	x_{34}
6	4	11	11	9	x_{35}
7	12	5	10	12	x_{47}
8	1	0	0	0	x_{56}
9	3	11	11	9	x_{57}
10	10	9	4	4	x_{68}
11	0	1	6	0	x_{78}

Set p = 1 in Equation 5.18 and using the above obtained ideal and anti-ideal solutions, the following auxiliary optimization problem is obtained:

$$\begin{aligned}\min h(x) &= (f1(x^+) - f1(x))\,/\,(f1(x^+) - f3(x^-)) + (f2(x^+) - f2(x))\,/\,(f2(x^+) - f4(x^-)) \\ &= 0.005772\,f1(x) + 0.008573\,f2(x) - 2.87901\end{aligned} \qquad (8.13)$$

subject to:

$$(f1 - 0.5\,x_{21} - 5\,x_{26} - 1.5\,x_{14} - 0.5\,x_{34} - 3\,x_{35} - 4\,x_{47} - 1.5\,x_{56} - 6\,x_{57} - 7\,x_{68} - 8\,x_{78}) \,/\, (0.5\,x_{21} + x_{26} + 0.5\,x_{14} + 0.5\,x_{34} + x_{35} + x_{47} + x_{68} + x_{78}) \geq \alpha$$

$$0.5\,x_{21} + 5\,x_{26} + 1.5\,x_{14} + 0.5\,x_{34} + 3\,x_{35} + 4\,x_{47} + 1.5\,x_{56} + 6\,x_{57} + 7\,x_{68} + 8\,x_{78} \leq f1$$

$$f1 \leq x_{21} + 6\,x_{26} + 2\,x_{14} + x_{34} + 4\,x_{35} + 5\,x_{47} + 2\,x_{56} + 7\,x_{57} + 8\,x_{68} + 9\,x_{78}$$

$$(f2 - 2\,x_{21} - x_{23} - 5\,x_{26} - 2\,x_{14} - 1.2\,x_{34} - 1.5\,x_{35} - 6\,x_{47} - x_{56} - x_{57} - 2\,x_{68} - 2\,x_{78}) \,/\, (x_{23} + x_{26} + 0.8\,x_{34} + 0.5\,x_{35} + x_{47} + x_{56} + 0.5\,x_{68} + 0.2\,x_{78}) \geq \alpha$$

$$2\,x_{21} + x_{23} + 5\,x_{26} + 2\,x_{14} + 1.2\,x_{34} + 1.5\,x_{35} + 6\,x_{47} + x_{56} + x_{57} + 2\,x_{68} + 2\,x_{78} \leq f2$$

$$f2 \leq 2\,x_{21} + 2\,x_{23} + 6\,x_{26} + 2\,x_{14} + 2\,x_{34} + 2\,x_{35} + 7\,x_{47} + 2\,x_{56} + 2\,x_{57} + 2.5\,x_{68} + 2.2\,x_{78}$$

$$\mathrm{x} \in \mathrm{X}$$

With a given value for the cut-off parameter, $\alpha = 0.5$, the above auxiliary problem becomes a linear programming problem, which can be solved by easily by the LINGO mixed-integer code [1992]. The optimal solution is:

$$x_{26} = 9,\ x_{14} = 10,\ x_{35} = 11,\ x_{47} = 5,\ x_{56} = 1,\ x_{57} = 10,\ x_{68} = 10,$$
$$x_{23} = 11,\ x_{34} = x_{78} = x_{21} = 0,\ f1 = 269.75,\ f2 = 176.75$$

This solution is a compromise between the minimum cost and the minimum time. The problem was also solved with the distance parameter $p = \infty$. The solution is:

$$x_{26} = 9,\ x_{14} = 10,\ x_{34} = 3,\ x_{35} = 8,\ x_{47} = 8,\ x_{56} = 1,\ x_{57} = 7,$$
$$x_{68} = 10,\ x_{23} = 11,\ x_{21} = x_{78} = 0,\ f1 = 255.5,\ f2 = 191.3.$$

Both above solutions are located at the non-dominated boundary.

8.4 Possibility Approach to Multi-level Minimum-cost Flow Problem

The above problem is extended to the multi-level or hierarchical decision-making problem. As has been discussed in previous chapters, a multi-level problem is a decentralized planning problem, where multiple agents are located at predominately hierarchical levels with mutual interactions. The system explicitly assigns each agent a unique objective and a set of decision variables as well as a set of common constraints that affect all agents. In addition to the many real-world applications, multi-level programming has also been applied to network design problems [Kim 1990, LeBlanc and Boyce 1986]. It is clear that the concept of multi-level can be applied to many other network problems.

Following Equation 6.22, a multi-level MCF problem can be represented as:

$$\min/\max\ f^1(x) = \sum_{(i,j)\in A} c^1{}_{ij} x_{ij} \qquad (1^{st}\text{ level}) \quad (8.14)$$

where $x^2{}_{ij}, x^3{}_{ij}, \ldots, x^k{}_{ij}$ solve:

$$\min/\max\ f^2(x) = \sum_{(i,j)\in A} c^2{}_{ij} x_{ij} \qquad (2^{nd}\text{ level})$$

where $x^3{}_{ij}, x^4{}_{ij}, \ldots, x^k{}_{ij}$ solve:

...

$$\min/\max\ f^k(x) = \sum_{(i,j)\in A} c^k{}_{ij} x_{ij} \qquad (k^{th}\text{ level})$$

subject to:

$$\sum_{(i,j)\in A} x_{ij} - \sum_{(i,j)\in A} x_{ij} = b(i),\ \forall i \in N$$
$$\ell_{ij} \le x_{ij} \le u_{ij},\ \forall (i,j) \in A$$
$$x_{ij} \ge 0,\ \forall (i,j) \in A$$

where all x_{ij} are integers, $x_{ij} = (x^1{}_{ij}, x^2{}_{ij}, \ldots, x^k{}_{ij})$, $i = 1, \ldots, m$, $j = 1, \ldots, n$ and $k = 1, \ldots, K$..

The multi-level programming problem represented by Equation 8.14 is classified as a NP-hard problem. Thus it is difficult to solve. As discussed in the previous chapters, to partly overcome this difficulty, a supervised search procedure in the fuzzy domain was used, which offers a convenient tool to manage this type of problem. This procedure is based on the concept of tolerance membership function and multi-objective optimization to handle the goals and the decisions. Since the MCF problem is a minimization problem, Equation 6.4 for the goal will be modified as:

$$\mu_{f1}(f^1(x)) = \begin{cases} 1, \text{ if } f^1(x) < f^{1\text{'}} \\ \dfrac{f^{1U}(x) - f^1(x)}{f^{1U} - f^{1\text{'}}}, \text{ if } f^{1\text{'}} \le f^1(x) \le f^{1U} \\ 0, \text{ if } f^1(x) > f^{1U} \end{cases} \qquad (8.15)$$

where $f^{1\text{'}}$ and f^{1U} are the acceptable range of the goal for the upper-level DM. The upper-level DM also sets the acceptable range of her decisions in terms of membership function, which is similar to Equation 6.3:

$$\mu_{xij}(x_{ij}) = \begin{cases} \dfrac{x_{ij} - (x_{ij}^{U} - p_{ij})}{p_{ij}}, \text{ if } x_{ij}^{U} - p_{ij} \le x_{ij} \le x_{ij}^{U} \\ \dfrac{(x_{ij}^{U} + p_{ij}) - x_{ij}}{p_{ij}}, \text{ if } x_{ij}^{U} < x_{ij} \le x_{ij}^{U} + p_{ij} \\ 0, \text{ otherwise} \end{cases} \tag{8.16}$$

where p_{ij} is the two-sided tolerance for the decision variable x_{ij}

This fuzzy information for the upper level restricts the lower-level DM's feasible space. For each possible solution of the upper-level DM, the lower-level DM also defines his goal as the following membership function:

$$\mu_{f2}(f^2(x)) = \begin{cases} 1, \text{ if } f^2(x) < f^{2\cdot} \\ \dfrac{f^{2L}(x) - f^2(x)}{f^{2L} - f^{2\cdot}}, \text{ if } f^{2\cdot} \le f^2(x) \le f^{\ L} \\ 0, \text{ if } f^2(x) > f^{2L} \end{cases} \tag{8.17}$$

where $f^{2\cdot}$ and f^{2L} are the acceptable range of the goal for the lower-level DM.

Using Equations 8.15, 8.16 and 8.17, the auxiliary equation for the fuzzy MCF multi-level problem can be established. The procedure for establishing this auxiliary equation involves two steps, The first step is to represent the imprecise parameters by the use of the possibility concept such as Equation 5.51. The second step involves the processing of the multi-level structure, which can be done in the same fashion as that discussed in Chapter 6.

Using the possibility concept to represent the imprecise parameters, which are trapezoidal fuzzy numbers, we obtain the following auxiliary equation:

$$\text{min/max} f^1 (x^1_{ij}, x^2_{ij}) \tag{8.18}$$

where x^2_{ij} solves

$$\text{min/max} f^2 (x^1_{ij}, x^2_{ij})$$

subject to:

$$\alpha\,(l_{ij2} - l_{ij1}) + l_{ij1} \le x_{ij} \le u_{ij4} - \alpha\,(u_{ij2} - u_{ij1}),\ \forall\ i \text{ and } j$$
$$\textstyle\sum_i \sum_j c_{ij1} x_{ij} \le f^1 \le \sum_i \sum_j c_{ij1}\, x_{ij}, \text{ and } \theta^1_1 \ge \alpha$$
$$\textstyle\sum_i \sum_j c_{ij3} x_{ij} \le f^1 \le \sum_i \sum_j c_{ij4} x_{ij}, \text{ and } \theta^1_2 \ge \alpha$$
$$\textstyle\sum_i \sum_j c_{ij1} x_{ij} \le f^2 \le \sum_i \sum_j c_{ij1} x_{ij}, \text{ and } \theta^2_1 \ge \alpha$$
$$\textstyle\sum_i \sum_j c_{ij3} x_{ij} \le f^2 \le \sum_i \sum_j c_{ij4}\, x_{ij}, \text{ and } \theta^2_2 \ge \alpha$$
$$\alpha \in [0, 1]$$
$$x_{ij} \ge 0$$

where $x_{ij} = (x^1_{ij}, x^2_{ij})$, $i = 1, \ldots, m$, $j = 1, \ldots, n$, and all the x_{ij} are integers. It should be noted that the objectives consist of two decision variable sets as follows:

$$f^1 = \Sigma_i\Sigma_j\, c^{11}_{ij}x^1_{ij} + c^{12}_{ij}x^2_{ij}$$
$$f^2 = \Sigma_i\Sigma_j\, c^{21}_{ij}x^1_{ij} + c^{22}_{ij}x^2_{ij}.$$

For simplification in notation, the constraints of Equation 8.18 will be represented by $x \in X_2$.

The next step is to optimize the total degree of satisfaction for the bi-level system. The upper-level DM will set his goal and decisions as fuzzy membership functions by the use of Equations 8.15 and 8.16, and transfer them to the constraints of the lower-level DM. At the same time, the lower-level DM is also establishing her goal as a fuzzy membership function such as Equation 8.17. Now, Equation 8.18 can be reformulated into the following:

$$\text{min/max}\, f^2(x^1_{ij}, x^2_{ij}) \qquad (8.19)$$

subject to:

$$\begin{aligned}
&(f^1(x_{ij}^+) - f^1)/(f^1(x_{ij}^+) - f^{1\prime}) \geq \lambda^1 \\
&[(x^{1\,U}_{ij} + p_{ij}) - x^1_{ij}]/p_{ij} \geq \lambda^2_{ij},\ \forall\, i \text{ and } j \\
&[x^1_{ij} - (x^{1\,U}_{ij} - p_{ij})]/p_{ij} \geq \lambda^2_{ij},\ \forall\, i \text{ and } j \\
&(f^2(x_{ij}^+) - f^2)/(f^2(x_{ij}^+) - f^{2\prime}) \geq \lambda^3 \\
&((f^2 - f^{2\prime})/(f^2(x_{ij}^+) - f^{2\prime}) \geq \lambda^3 \\
&\lambda^1, \lambda^2, \lambda^3, \alpha, \theta^1_1, \theta^2_1 \in [0, 1],\ \forall\, i \text{ and } j \\
&\theta^1_2, \theta^2_2 \in [0, 1] \\
&x \in X_2
\end{aligned}$$

where $x_{ij} = (x^1_{ij}, x^2_{ij}) \geq \mathbf{0}$, $i = 1, \ldots, m$, $j = 1, \ldots, n$. Again, the objectives are composed of two decision variable sets as follows:

$$f^1 = \Sigma_i\Sigma_j\, c^{11}_{ij}x^1_{ij} + c^{12}_{ij}x^2_{ij}$$
$$f^2 = \Sigma_i\Sigma_j\, c^{21}_{ij}x^1_{ij} + c^{22}_{ij}x^2_{ij}$$

Using the same procedure that was used to obtain Equation 6.13, we can transform the above equation into the final desired programming problem. In other words, in order to satisfy all the DMs involved, we must have

$$\lambda = \min\{\lambda^1, \lambda^2_{ij}, \lambda^3\} \qquad (8.20)$$

for $i = 1, \ldots, m$, $j = 1, \ldots, n$.. Follow the same procedure that was used to obtain Equation 6.13, we finally have:

$$\max \lambda \qquad (8.21)$$

subject to:

$$(f^1(x_{ij}^+) - f^1)/(f^1(x_{ij}^+) - f^{1\prime}) \geq \lambda$$

$$[(x^{1}{}_{ij}{}^{U} + p_{ij}) - x^{1}{}_{ij}] / p_{ij} \geq \lambda, \ \forall \ i \text{ and } j$$
$$[x^{1}{}_{ij} - (x^{1}{}_{ij}{}^{U} - p_{ij})] / p_{ij} \geq \lambda, \ \forall \ i \text{ and } j$$
$$(f^{2}(x_{ij}{}^{+}) - f^{2}) / (f^{2}(x_{ij}{}^{+}) - f^{2\prime}) \geq \lambda$$
$$((f^{2} - f^{2\prime}) / (f^{2}(x_{ij}{}^{+}) - f^{2\prime}) \geq \lambda$$
$$\mathrm{x} \in \mathrm{X}_2.$$
$$\lambda, \alpha, \theta^{1}{}_{1}, \theta^{2}{}_{1} \in [0, 1]$$
$$\theta^{1}{}_{2}, \theta^{2}{}_{2} \in [0, 1]$$

where $x_{ij} = (x^{1}{}_{ij}, x^{2}{}_{ij}) \geq \mathbf{0}$, $i = 1, ..., m$, and $j = 1, ..., n$. The objectives are composed of two decision variable sets as follows:

$$f^{1} = \Sigma_i\Sigma_j \, c^{11}{}_{ij} \, x^{1}{}_{ij} + c^{12}{}_{ij} \, x^{2}{}_{ij}$$
$$f^{2} = \Sigma_i\Sigma_j \, c^{21}{}_{ij} \, x^{1}{}_{ij} + c^{22}{}_{ij} \, x^{2}{}_{ij}$$

for all i and j.

The maximum degree of satisfaction can be obtained by using Equation 8.21. However, this possibilistic expression is non-linear. The problem reduces to linear form if the DM provides a fixed value of cut-off parameter $\alpha \in [0, 1]$.

Example 8.3. Consider the same two-level decision-making problem solved in Example 8.2, except the two objectives are treated as two separate levels. The problem is to minimize the total cost f^{1} for the upper-level DM and the passing time f^{2} for the lower-level DM. Using the information in Table 8.3, the fuzzy ranges for the objectives are: $f^{1} \in [236.5, 409.75]$ and $f^{2} \in [176.6, 293.25]$. Assume that the upper-level DM has two control variables. The value of the first control decision, x_{14}, is around 10 with negative and positive side tolerances of 6 and 5, respectively. The second, x_{35}, is around 4 with negative and positive side tolerances of 4 and 7, respectively. The additional constraints are given as:

$$x_{14} - 4 \geq 6\lambda^{2}{}_{14}$$
$$15 - x_{14} \geq \lambda^{2}{}_{14}$$
$$x_{35} \geq 7\lambda^{2}{}_{35}$$

The two DMs set their tolerance on their goals as:

$$409.75 - f^{1} \geq 173.25 \, \lambda^{1}$$
$$293.25 - f^{2} \geq 116.65 \, \lambda^{3}$$

The decision must satisfy all the DMs, or, in other words, Equation 8.20 for the present problem must take the following form:

$$\lambda = \min\{\lambda^{1}, \lambda^{2}{}_{14}, \lambda^{2}{}_{35}, \lambda^{3}\}$$

Substituting the above given information into Equation 8.21, the auxiliary problem becomes:

max λ

subject to:

$$
\begin{aligned}
&15 - x_{14} \geq 5\,\lambda \\
&x_{14} - 4 \geq 6\,\lambda \\
&11 - x_{35} \geq 7\,\lambda \\
&x_{35} \geq 4\,\lambda \\
&409.75 - f^1 \geq 173.25\,\lambda \\
&239.25 - f^2 \geq 173.25\,\lambda \\
&(f^1 - 0.5\,x_{21} - 5\,x_{26} - 1.5\,x_{14} - 0.5\,x_{34} - 3\,x_{35} - 4\,x_{47} - 1.5\,x_{56} - 6 \\
&x_{57} - 7\,x_{68} - 8\,x_{78})\,/\,(0.5\,x_{21} + x_{26} + 0.5\,x_{14} + 0.5\,x_{34} + x_{35} + x_{47} + \\
&x_{68} + x_{78}) \geq \alpha \\
&0.5\,x_{21} + 5\,x_{26} + 1.5\,x_{14} + 0.5\,x_{34} + 3\,x_{35} + 4\,x_{47} + 1.5\,x_{56} + 6\,x_{57} + \\
&7\,x_{68} + 8\,x_{78} \leq f^1 \\
&f^1 \leq x_{21} + 6\,x_{26} + 2\,x_{14} + x_{34} + 4\,x_{35} + 5\,x_{47} + 2\,x_{56} + 7\,x_{57} + 8\,x_{68} + \\
&9\,x_{78} \\
&(f^2 - 2\,x_{21} - x_{23} - 5\,x_{26} - 2\,x_{14} - 1.2\,x_{34} - 1.5\,x_{35} - 6\,x_{47} - x_{56} - x_{57} \\
&- 2\,x_{68} - 2\,x_{78})\,/\,(x_{23} + x_{26} + 0.8\,x_{34} + 0.5\,x_{35} + x_{47} + x_{56} + 0.5\,x_{68} \\
&+ 0.2\,x_{78}) \geq \alpha \\
&x_{21} + x_{23} + 5\,x_{26} + 2\,x_{14} + 1.2\,x_{34} + 1.5\,x_{35} + 6\,x_{47} + x_{56} + x_{57} + 2\,x_{68} \\
&+ 2\,x_{78} \leq f^2 \\
&f^2 \leq 2\,x_{21} + 2\,x_{23} + 6\,x_{26} + 2\,x_{14} + 2\,x_{34} + 2\,x_{35} + 7\,x_{47} + 2\,x_{56} + 2 \\
&\qquad x_{57} + 2.5\,x_{68} + 2.2\,x_{78} \\
&x_{14} - x_{21} = 10 \\
&x_{21} + x_{23} + x_{26} = 20 \\
&x_{34} + x_{35} - x_{23} = 0 \\
&x_{47} - x_{14} - x_{34} = -5 \\
&x_{56} + x_{57} - x_{35} = 0 \\
&x_{68} - x_{56} - x_{26} = 0 \\
&x_{78} - x_{47} - x_{57} = -15 \\
&- x_{68} - x_{78} = -10 \\
&x_{21} \leq 11 - 2\,\alpha \\
&\alpha \leq x_{23} \leq 13 - 4\,\alpha \\
&x_{26} \leq 11 - 2\,\alpha \\
&2\,\alpha \leq x_{14} \leq 16 - \alpha \\
&x_{34} \leq 9.5 - 4.5\,\alpha \\
&x_{35} \leq 12 - 2\,\alpha \\
&2\,\alpha \leq x_{47} \leq 14.5 - 4.5\,\alpha \\
&x_{56} \leq 22 - 2\,\alpha \\
&x_{57} \leq 17 - 2\,\alpha \\
&\alpha \leq x_{68} \leq 12 - 2\,\alpha \\
&x_{78} \leq 16.5 - 1.5\,\alpha \\
&\lambda, \alpha \in [0, 1]
\end{aligned}
$$

where all the x_{ij} are positive integers. For simplification in notation, the constraints for the above problem except the first 6 inequalities will be denoted as $x \in X_3$. In other words, the expression $x \in X_3$ represents the constraint set of the above problem except it doesn't include the first 6 inequalities.

Set $\alpha = 0.5$, this auxiliary integer programming problem was solved. The optimal degree of satisfaction obtained is 0.429. The other obtained values are:

$$f^1 = 260.5,\ f^2 = 187.2,$$
$$x_{14} = 10, x_{35} = 8, x_{26} = 10, x_{34} = 2, x_{47} = 7, x_{57} = 8, x_{68} = 10,$$
$$x_{23} = 10,\ x_{21} = x_{56} = x_{78} = 0$$

Instead of the non-compensatory *max-min* approach, which was first proposed by Zimmermann [1985] based on the seminal paper by Bellman and Zadeh [1970], the compensatory operator based on Werners' interpretation of the connectives *and* and *or* is used in the next example.

Example 8.4. Consider Example 8.3 again, except now the compensatory operator due to Werners is used in forming the auxiliary equation. Using the same approach as that used to obtain Equation 8.7, the following auxiliary equations were obtained for the network problem solved in Example 8.3:

$$\max \mu_{\text{and}} = \lambda + (1-\gamma)(\lambda^1 + \lambda^2_{14} + \lambda^2_{35} + \lambda^3)/4$$

subject to:

$$15 - x_{14} \geq 5\,(\lambda + \lambda^2_{14})$$
$$x_{14} - 4 \geq 6\,(\lambda + \lambda^2_{14})$$
$$11 - x_{35} \geq 7\,(\lambda + \lambda^2_{35})$$
$$x_{35} \geq 4\,(\lambda + \lambda^2_{35})$$
$$409.75 - f^1 \geq 173.25\,(\lambda + \lambda^1)$$
$$239.25 - f^2 \geq 173.25\,(\lambda + \lambda^3)$$
$$x \in X_3$$

where all the x_{ij} are positive integers and $x \in X_3$ represents the remaining constraints, which are the same as that in the auxiliary equation of Example 8.3. The expression $x \in X_3$ does not include the first 6 inequalities in the auxiliary equation of Example 8.3.

With a given value for the cut-off parameter $\alpha = 0.5$, this problem was solved. The results obtained are:

$$f^* = (f^{1*}, f^{2*}) = (255.75, 192.05),$$
$$x_{14} = 10, x_{35} = 7, x_{26} = 10, x_{34} = 3, x_{47} = 8, x_{57} = 7, x_{68} = 10,$$
$$x_{23} = 10, x_{21} = x_{56} = x_{78} = 0,$$
$$\mu_{\text{and}} = 0.560, \gamma = 0.5$$

To investigate the effect of compensation, eleven cases with the compensatory parameter γ varies from 0 to 1.0 and with α =0.5. We find the results are fairly stable, and the solution stays the same for $\gamma \geq 0.6$.

8.5 Discussions

To avoid the difficulties encountered in using the possibility approach, the concept of fuzzy tolerance in the form of cut-off parameter was proposed to replace the possibility exceedence operator. [Dubois and prade 1988]. The cut-off parameter is similar to subjective possibility and the DM can specify the value of the cut-off parameter in consideration of the actual problem involved. The added advantage is that a one third reduction in the number of capacity constraints resulted. This reduction provides great benefits for large-scale problems.

The proposed approach uses the advantages of both the possibility concept and Zimmermann's preference-based model. The possibility concept is used to represent or to model the vagueness of the imprecise parameters and the preference-based model is used for simplifying the multi-level structure.

Most examples in this chapter are solved by the LINGO mixed-integer code [1992] However, this code might be inefficient for dealing with large-scale problems. Some of the specialized network software may be used to solve large problems. However, some of these codes can only handle all-integer problem and cannot solve the mixed-integer fuzzy MCF problem. Although we can process the non-integer variables by scaling, for example, by multiply all the values by 1000 to make the variables integers [Kennington and Helgason 1980], it is not a good method to handle fuzzy problems. Thus, there is a need for software for solving the fuzzy MCF type of problems.

References

Ahuja, R.K., T.L. Magnanti and J.B. Orlin, *Network Flows*. Prentice-Hall, Englewood, NJ (1993).

Aiyoshi, E. and K. Shimizu, Hierarchical decentralized systems and its new solution by a barrier method. *IEEE Transactions on Systems, Man, and Cybernetics*. **11(6)**, 444-449 (1981a).

Aiyoshi, E. and K. Shimizu, A new computational method for Stackelberg and min-max problems by use of a penalty method. *IEEE Transactions on Automatic Control*. 26(2), 460-466 (1981b).

Aiyoshi, E. and K. Shimizu, A solution method for the static constrained Stackelberg problem via penalty method. *IEEE Transactions on Automatic Control*. **29(12)**, 1111-1114 (1984).

Aksoy, Y., Interactive Multiple objective decision making: a bibliography (1965-1988). *Management Research News* **13**, 1-8, (1990).

Al-Khayyal, F.A., An implicit enumeration procedure for the general linear complementarity problem, *Math. Progr. Studies*, 31, 1-20 (1987).

Anandalingam, G., A mathematical programming model of decentralized multi-level systems. *J. of Operational Research Society*. **39**, 1021-1033 (1988).

Anandalingam, G. and V. Apprey, Multi-level programming and conflict resolution. *European J. of Operational Research* **51**, 233-247 (1991).

Anandalingam, G. and T.L. Friesz, Hierarchical Optimization: An Introduction. *Annals of Operations Research*. (ed. by G. Anandalingam and T.L. Friesz) **34**, 1-11 (1992).

Anandalingam, G. and D.J. White, A solution for the linear static Stackelberg problem using penalty functions. *IEEE Transaction on Automatic Control* **35**, 1170-1173 (1990).

Apprey, V., *Multi-agent Mathematical Programming with an Application to Conflict Resolution*. Ph.D. dissertation, University of Virginia, Charlottesville, VA (1987).

Aubin, J.-P., *Optima and Equilibria: An Introduction to Nonlinear Analysis*. Springer-Verlag, Berlin (1993).

Audet, C., P. Hansen, B. Jaumard and G. Savard, Links between linear bilevel and mixed 0-1 programming problems. *J. of Optimization Theory and Applications* **93(2)**, 273-300 (1997).

Bard, J.F., A grid search algorithm for the linear bilevel programming problem. in *14th Annual Meeting of the American Institute for Decision Sciences*, San Francisco, CA, Nov. 22-24, 1982, vol. 2, 256-258 (1982).

Bard, J.F., Coordination of a multidivisional organization through two levels of management. *Omega* **11**, 457-468 (1983a).

Bard, J.F., An efficient point algorithm for a linear two-stage optimization problem. *Operations Research* **31**, 670-684 (1983b).

Bard, J.F., An algorithm for solving the general bilevel programming problems, *Math. Operations Research*, 8, 260-272 (1983).

Bard, J.F., Optimality conditions for the bilevel programming problem. *Naval Research Logistic Quarterly* **31**, 13-26 (1984a).

Bard, J.F., An investigation of the linear three level programming problem, *IEEE Transactions on Systems, Man, and Cybernetics* **14(5)**, 711-717 (1984b).

Bard, J.F., Convex two-level optimization, *Mathematical Programming*, 40,15-27 (1988).

Bard, J.F., Some properties of the bilevel programming problem, *J. Optimization Theory and Applications*, 68, 371-378 (1991).

Bard, J.F. and J.E. Falk, An explicit solution to the multi-level programming problem. *Computers and Operations Research* **9**, 77-100 (1982).

Bard, J.F. and J.T. Moore, A branch and bound algorithm for the bilevel programming problem. *SIAM J. on Scientific and Statistical Computing* **11**, 281-292 (1990).

Bard, J.F. and J.T. Moore, An algorithm for the discrete bilevel programming problem. *Naval Research Logistics* **39**, 419-435 (1992).

Basar, T. and G. J. Olsder, *Dynamic Noncooperative Game Theory*. Academic Press, New York (1982).

Beale, E.M.L., and R.E. Small, Mixed integer programming by branch-and-bound technique, Proceedings of the third IFIP Congress, Vol.2, 450-451 (1965).

Bellman, R.E. and L.A. Zadeh, Decision making in a fuzzy environment. *Management Science* **17**, B141-164 (1970)

Bellman, R.E and M.Giertz, On the analytic formalism of the theory of fuzzy sets, *Information Science*, *5*, 149-157 (1973).

Ben-Ayed, O. and C. E. Blair, Computational difficulties of bilevel linear programming. *Operations Research* **38**, 556-560 (1990).

Ben-Ayed, O., Bilevel linear programming. *Computers and Operations Research* **20**, 485-501 (1993).

Benson, H.P.,On the structure and properties of a linear multievel programming problem. *J of Optimization Theory and pplications* **60(3)**, 353-373 (1989).

Bialas, W.F. and M.H. Karwan, On two-level optimization. *IEEE Transaction on Automatic Control* **AC-27**, 211-214 (1982).

Bialas, W.F. and M.H. Karwan, Two-level linear programming. *Management Science* **30**, 1004-1020 (1984).

Bit, A.K., M.P. Biswal and S.S. Alam, Fuzzy programming approach to multicriteria decision making transportation problem. *Fuzzy Sets and Systems* **50(2)**, 135-141 (1992).

Bit, A.K., M.P. Biswal and S.S. Alam, Fuzzy programming approach to multiobjective solid transportation problem. *Fuzzy Sets and Systems* **57(2)**, 183-194 (1993a).

Bit, A.K., M.P. Biswal and S.S. Alam, An additive fuzzy programming model for multiobjective transportation problem. *Fuzzy Sets and Systems* **57(3)**, 313-319 (1993b).

Bracken, J. and J.M. McGill, Mathematical programs with optimization problems in the constraints. *Operations Research* **21**, 37-44 (1973).

Bracken, J. and J.M. McGill, A method for solving mathematical programs with nonlinear programs in the constraints. *Operations Research* **22**, 1094-1101 (1974).

Breiner, A. and M. Avriel, On the solution of quantitative policy analysis models using bi-level programming. *Socio-Economic Planning Sciences* **28(3)**,179-195 (1994).

Brickley, J.A., C.W. Smith, and J.L. Zimmerman, *Organizational Architecture: a managerial economics approach*. Irwin Inc., NY (1996).

Buckley, J.J., Possibilistic linear programming with triangular fuzzy numbers. *Fuzzy Sets and Systems* **20(1)**, 135-138 (1988).

Buckley, J.J., Solving possibilistic linear programming problems. *Fuzzy Sets and Systems* **31(3)**, 329-341 (1989).

Burton, R.M., The multilevel approach to organizational issues of the firm - a critical review. *Omega* **5**, 395-414 (1977).

Calamai, P.H. and L.N. Vicente, Generating quadratic bilevel programming test probems. *ACM Transactions on Mathematical Software* **20(1)**, 103-119 (1994).

Calvete, H.I. and C. Galé, The bilevel linear/linear fractional programming problem. *European J. of Operational Research* **114(1)**, 188-197 (1999).

Candler, W., A linear bilevel programming algorithm: a comment. *Computers and Operations Research* **15**, 297-298 (1988).

Candler, W., J. Fortuny-Amat, and B. McCarl, The potential role multilevel programming in agricultural economics. *American J. of Agricultural Economics* **63(3)**, 521-531 (1981).

Candler, W. and R. Townsley, A linear two-level programming problem. *Computers and Operations Research* **9**, 59-76 (1982).

Carlsson, C. and P. Korhonen, A parametric approach to fuzzy linear programming. *Fuzzy Sets and Systems* **20**, 17-30 (1986).

Cassidy, R., M. Kirby and W. Raike, Efficient distribution of resounrces through three levels of government, *Mgmt. Sci.*, *17*, B462-473 (1971).

Chanas, S., Fuzzy optimization in networks. in *Optimization Models Using Fuzzy Sets and Possibility Theory* (eds. by J. Kacprzyk and S.A. Orlovski), D. Reidel, Dordrecht, 303-327 (1987).

Chanas, S., M. Delgado, J.L. Verdegay and M.A. Vila, Fuzzy optimal flow on imprecise structures. *European J. Operational Research* **83(3)**, 568-580 (1995).

Chanas, S. and W. Kołodziejczyk, Maximum flow in a network with fuzzy arc capacities. *Fuzzy Sets and Systems* **8(2)**, 165-173 (1982).

Chanas, S. and W. Kołodziejczyk, Real-values flows in a network with fuzzy arc capacities. *Fuzzy Sets and Systems* **13(2)**, 139-151 (1984).

Chanas, S. and W. Kołodziejczyk, Integer flows in networks with fuzzy capacity constraints. *Networks* **16(1)**, 17-31 (1986).

Chanas, S., W. Kołodziejczyk and A. Machaj, A fuzzy approach to the transportation problem. *Fuzzy Sets and Systems* **13(3)**, 211-221 (1984).

Chang, P.T and E.S. Lee, Fuzzy decision making: a survey, in*Between Mind and Computer: Fuzzy Science and Engineering*, P.Z.Wang, K.Floe (eds), World Scientific Publishing Co.,Singapore, 139-182 (1993)

Chang, P.T. and E..S. Lee, Ranking fuzzy sets based on the concept of existence, *Computers Mathematics Applications*, **27(9/10)**,1-21 (1994)

Charnes,A and W.W. Cooper, Goal programming and multi-objective optimization, *European Journal Operations Research*, **1**,39-54 (1977)

Chen, S.J. and C.L. Hwang, *Fuzzy Multiple Attribute Decision Making: Methods and Applications*. Springer-Verlag, Berlin (1992).

Cheng, C.B. and E.S.Lee, " Response surface based on fuzzy adaptvie network, *Int. Journal of Fuzzy Systems*, **1**, 14-24 (1999)

Clark, P.A. and A.W. Westerberg, A note on the optimality conditions for the bilevel programming problem. *Naval Research Logistic Quarterly* **35**, 413-418 (1988).

Davis, L., (ed), *Handbook of Genetic Algorithms*, Nan Nostrand Reinhold, New York (1991).

de Silva, A.H., *Sensitivity Formulas for nonlinear factorable programming and Their Application to the Solution of an Implicitly Defined Model of U.S. Crude Oil Production*. D.Sci. thesis, George Washington University, Washington, D.C. (1978).

Delgado, M., J.L. Verdegay and M.A. Vila, Fuzzy transportation problems: a general analysis. in *Optimization Models Using Fuzzy Sets and Possibility Theory* (eds. by J. Kacprzyk and S.A. Orlovski), D. Reidel, Dordrecht, 342-358 (1987).

Dempe, S., A necessary and a sufficient optimality condition for bilevel programming problems. *Optimization* **25**, 341-354 (1992).

Dempe, S., First-order necessary optimality conditions for general bilevel programming problems. *J. of Optimization Theory and Applications* **95(3)**, 735-739 (1997).

Dempe, S. and H. Schmidt, On an algorithm solving two-level programming problems with nonunique lower level solution. *Computational Optimization and Applications* **6**, 227-249 (1996).

Dubois, D. and H. Prade, *Fuzzy Sets and Systems*. Academic Press, NY (1980).

Dubois, D. and H. Prade, The Use of Fuzzy Numbers in Decision Analysis. in *Fuzzy Information and Decision Processes*, eds. by M.M. Gupta and E. Sanchez, 309-321, North-Holland Publishing Company, Holland, (1982).

Dubois, D. and H. Prade, Criteria aggregation and ranking of alternatives in the framework of fuzzy set theory. in *TIMS/Studies in the Management Sciences* **20**, (ed. by H.-J. Zimmermann, L.A. Zadeh, and B.R. Gaines) North-Holland, Amsterdam, 209-240 (1984).

Dubois, D. and H. Prade, A review of fuzzy set aggregation connectives. *Information Science* **36**, 85-121 (1985).

Dubois, D. and H. Prade, *Possibility Theory*, Plenum Press, New York (1988)

Dyckhoff, H., Basic concepts for a theory of evaluation: Hierarchical aggregation via autodistributive connectives in fuzzy set theory. *European J. Operational Research* **20**, 221-233 (1985).

Dyckhoff, H. and W. Pedrycz, Generalized means as model of compensative connectives. *Fuzzy Sets and Systems* **14**, 143-154 (1984).

Edmunds, T.A. and J.F. Bard, Algorithms for nonlinear bilevel mathematical programs. *IEEE Transactions on Systems, Man, and Cybernetics* **21(1)**, 83-89 (1991).

Evans, G.W., An overview of techniques for solving multiobjective mathematical programming. *Management Science* **30(11)**, 1268-1282 (1984).

Falk, J.E. and Jiming Liu, On bilevel programming, part I: general nonlinear cases. *Mathematical Programming* **70(1)**, 47-72 (1995).

Fedrizzi, M., J. Kacprzyk and M. Roubens (eds.), *Interactive Fuzzy Optimization*. Springer-Verlag, Berlin, (1991).

Fortuny-Amat, J. and B. McCarl, A representation and economic interpretation of a two-level programming problem. *J. of Operational Research Society* **32**, 783-792 (1981).

Frangioni, A., On a new class of bilevel programming problems and its use for reformulating mixed integer problems. *European J. of Operational Research* **82**, 615-646 (1995).

Friesz, T.L., R.L. Tobin, H.J. Cho, and N.J. Mehta, Sensitivity analysis based heuristic algorithms for mathematical programs with variational inequality constraints. *Mathematical Programming* **48**, 265-284 (1990).

Fung, L.W. and K.S. Fu, An axiomatic approach to rational decision making in a fuzzy environment. in *Fuzzy Sets and Their Applications to Cognitive and Decision Processes* (ed. by L.A. Zadeh, K.S. Fu, K. Tanaka, and M. Shimura), Academic Press, NY, 227-256 (1975).

Gendreau, M., P. Marcotte and G. Savard, A hybrid Tabu-ascent algorithm for the linear bilevel programming problem. *J. of Global Optimization* **8(3)**, 217-233 (1996).

Geoffrion, A.M. and W.W. Hogan, Coordination of two-level organizations with multiple objectives. in *Techniques of Optimization* (ed. by A.V. Balakrishnan), Academic Press, NY, 455-466 (1972).

Goldberg, D.E., *Genitic Algorithms in Search Optimization, and Machine learning*, Addison-Wesley, Reading, MA (1989).

Hansen, P., B. Jaumard and G. Savard, New branch-and bound rules for linear bilevel programming. *SIAM J. of Scientific and Statistical Computing* **13(5)**, 1194-1217 (1992).

Haurie, A., G. Savard and D. J. White, A note on: an efficient point algorithm for a linear two-stage optimization problem. *Operations Research* **38**, 553-555 (1990).

Hersh H.M. and A. Caramazza, A fuzzy set approach to modifiers and vagueness in natural language. *J. of Experimental Psychology: General* **105(3)**, 254-276 (1976).

Hogarth, R.M., *Judgement and Choice: The Psychology of Decision*, 2nd ed. John Wiley & Sons, Chichester (1987).

Holland, J.H., *Adaptation in neural and artificial systems*, University of Michigan Press, Ann Arbor, Michagan (1975).

Hwang, C.L. and A.S.M. Masud, *Multiple Objectives Decision Making: Methods and Applications*. Springer-Verlag, Berlin (1979).

Hwang, C.L. and K. Yoon, *Multiple Attribute Decision Making: Methods and Applications*. Springer-Verlag, Berlin (1981).

Hwang, C.L., S.R. Paidy, K. Yoon, and A.S.M. Masud, *Mathematical programming with multiple objectives: a tutorial. Computers and Operations Research* **7**, 5-31 (1980).

Ignizio, J.P. and S.C. Daniels, Fuzzy multicriteria integer programming via fuzzy generalized networks. *Fuzzy Sets and Systems* **10**, 261-270 (1982)

Ishizuka, Y. and E. Aiyoshi, Double penalty method for bilevel optimization problems, *Annals of Operations Research*, 34, 73-88 (1992).

Jan, R.-H. and M.-S. Chern, Nonlinear interger bilevel programming. *European J. of Operational Research* **72,** 574-587 (1994).

Jeroslow, R.G., The polynomial hierarchy and a simple model for competitive analysis. *Mathematical Programming* **32,** 146-164 (1985).

Júdice, J.J. and A.M. Faustino, An experimental investigation of enumerative methods for the linear complementarity problem. *Computers and Operations Research* **15(5)**, 417-426 (1988).

Júdice, J.J. and A. Faustino, A sequential lcp method for bilevel linear programmig. *Annals of Operations Research* **34**, 89-106 (1992).

Júdice, J.J. and A. Faustino, The linear-quadratic bilevel programming problem. *INFOR* **32(2)**, 87-98 (1994).

Kacprzyk, J., *Multistage Decision-making under Fuzziness Theory and Applications*. Verlag TUV Rheinland, Koln (1983).

Karlof, J.K. and W. Wang, Bilevel programming applied to the flow shop scheduling problem. *Computers and Operations Research* **23, 443-451** (1996).

Kacara, M. Topology optimization with displacement constraints: a bilevel programming approach,*Structural Optimization*, 14, 256-263 (1997).

Kaufmann, A., *Introduction to the Theory of Fuzzy Subsets. Volume 1, Fundamental Theoretical Elements*,

Kennington, J.L. and R.V. Helgason, *Algorithms for Network Programming*. John Wiley, NY (1980). Academic Press, NY (1975).

Kickert, W. J. M., *Fuzzy Theories on Decision Making - A Critical Review*, Martinus Nijhoff, Leiden (1978).

Kim, T.J., Advanced *Transport and Spatial Systems Models*. Springer-Verlag, NY (1990).

Kim, T.J. and S. Suh, Toward development a national transportation planning model: A bi-level programming approach for Korea. *The Annals of Regional Science* **12**, 65-80 (1988).

Kirsch, U., Two-level optimization of prestressed structures. *Engineering Structures* **19(4)**, 309-317 (1997).

Klein, C.M., Fuzzy shortest paths. *Fuzzy Sets and Systems* **39(1)**, 27-41 (1991).

Klir, G.J. and T.A. Folger, *Fuzzy Sets, Uncertainty, and Information*. Prentice-Hall, Englewood, NJ (1988).

Klir, G.J. and B. Yuan, *Fuzzy Sets and Fuzzy logic: Theory and Application.*, Prentice-Hall, Englewood Cliffs, NJ (1995).

Kocvara, M., Topology optimization with displacement constraints: a bilevel programming approach. *Structural Optimization* **14(4)**, 256-263 (1997).

Kolstad, C.D. , *A review of the literature on bi-level mathematical programming*. Report no. LA-10234-MS, Los Alamos National Laboratory, Los Alamos, New Mexico (1985).

Kolstad, C.D. and L.S. Lasdon, Derivative evaluation and computational experience with large bilevel mathematical programs. *J. of Optimization Theory with Applications* **65(3)**, 485-499 (1990).

Korhonen, P., H. Moskowitz and J. Wallenius, Multiple criteria decision support: a review. *European J. of Operational Research* **63**, 361-375 (1992).

Krishnapuram, R., J. Lee, Fuzzy-connective-based hierarchical aggregation networks for decision making. *Fuzzy Sets and Systems* **46(1)**, 11-27 (1992).

Lai, Y.J., TOPSIS for MODM. *European J. of Operational Research* **76**, 486-500 (1994).

Lai, Y.J., IMOST: Interactive Multiple Objective System Technique. *J. of the Operational Research Society* **46**, 958-976, (1995).

Lai, Y.J., Hierarchical optimization: a satisfactory solution. *Fuzzy Sets and Systems* **77(3)**, 321-335 (1996).

Lai, Y.J. and C.L. Hwang, *Fuzzy Mathematical Programming: Methods and Applications*. Springer-Verlag, Berlin (1992).

Lai, Y.J. and C.L. Hwang, Possibilistic linear programming for managing interest rate risk. *Fuzzy Sets and Systems* **54**, 135-146 (1993).

Lai, Y.J. and C.L. Hwang, *Fuzzy Multiple Objective Decision Making - Methods and Applications*. Springer-Verlag, Berlin (1994).

LeBlanc, L.J. and D.E. Boyce, A bilevel programming algorithm for exact solution of the network design programming with user-optimal flows. *Transportation Research* **20B(3)**, 259-265 (1986).

Lee, E.S. and R.J. Li, Fuzzy multiple objective programming and compromise programming with Pareto optimum. *Fuzzy Sets and Systems* **53**, 275-288 (1993).

Lee, E. S. and Q. Zhu, *Fuzzy and Evidence Reasoning*, Physica-Verlag,Heidelberg, Germany (1995)

Leung, Y., Compromise programming under fuzziness. *Control and Cybernetics* **13(3)**, 203-214 (1984).

Li, R.J., *Multiple Objective Decision Making in a Fuzzy Environment*. Ph.D. Dissertation, Dept. of Industrial Engineering, Kansas State University, Manhattan, Kansas (1990).

Li, R.J. and E.S. Lee, Fuzzy approaches to multicriteria de Novo programs. *J. of Mathematical Analysis and Applications* **153(1)**, 97-111 (1990).

Liebman, J., L. Lasdon, L. Schrage, and A. Waren, *Modeling and Optimization with GINO*. The Scientific Press, South San Francisco, CA (1986).

Lignola, M.B. and J. Morgan, Stability of regularized bilevel programming problems. *J. of Optimization and Applications* **93(3)**, 575-596 (1997).

Lin, Y.C., *A Study of Goal Programming*. M.S. Thesis, Dept. of Industrial Engineering, Kansas State University, Manhattan, Kansas (1988).

LINGO, Lindo System Inc., Chicago (1992).

Liu, B., Stackelberg-Nash equilibrium for multilevel programming with multiple followers using genetic algorithms. *Computers Math. Applic.* **36(7)**,79-89(1998).

Luhandjula, M.K., Compensatory operators in fuzzy linear programming with multiple objectives. *Fuzzy Sets and Systems* **8(3)**, 245-252 (1982).

Luhandjula, M.K., On possibilistic linear programming. *Fuzzy Sets and Systems* **18(1)**, 15-30 (1986).

Macal, C.M. and A.P. Hurter, Dependence of bilevel mathematical programs on irrelevant constraints. *Comprters and Operations Research* **24(12)**, 1129-1140 (1997).

Magnanti, T.L. and R.T Wong, Network design and transportation planning: models and algorithms, *Transportation Science*, 1-55 (1994)

Marcotte, P., Network design problem with congestion effects: a case of bilevel programming. *Mathematical Programming* 34, 142-162 (1986).

Marcotte, P. and G. Savard, A note on the Pareto optimality of solutions to the linear bilevel programming problem. *Computers and Operations Research* **18**, 355-359 (1991).

Mathieu, R., L.Pittard and G. Anandalingam, Genetic algorithm based approach to bi-level linear programming, R.A.I.R.O. Recherche Operationnelle, 28, 1-21 (1994)

Mathur, K. and M.C. Puri, A bilevel bottleneck problem. *European J. of Operational Research* **86(2)**, 337-344 (1995a).

Mathur, K. and M.C. Puri, On bilevel fractional programming. *Optimization* **35**, 215-226 (1995b).

McCarl, B.A. and T.H. Spreen, Price endogenous mathematical programming as a tool for sector analysis. *American J. of Agricultural Economics* **62(1)**, 87-102 (1980).

Meuriot, E. and J.M. Gates, Fishing allocations and optimal fees: a single- and multilevel programming analysis. *American J. of Agricultural Economics* **65(4)**, 711-721 (1983).

Migdalas, A., Bilevel programming in traffic planning: models, methods and challenge. *J. of Global Optimization* **7**, 381-405 (1995).

Migdalas, A. and P.M. Pardalos, Editorial: Hierarchical and bilevel programming. *J. of Global Optimization (Special Issue: Hierarchical and Bilevel programming)* **8**, 209-215 (1996).

Migdalas, A., P.M. Pardalos and P. Varbrand (eds.), *Multilevel Optimization: Algorithms and Applications (Nonconvex Optimization and Its Applications, Vol. 20)*. Kluwer, Boston (1998).

Min, H. and J. Storbeck, On the origin and persistence of misconceptions in goal programming. *J. of Operational Research Society* **42(4)**, 301-312 (1991).

Mizumoto, M., Pictorial representations of fuzzy connectives, Part I: cases of t-norms, t-conorms and averaging operators. *Fuzzy Sets and Systems* **31(2)**, 217-242 (1989a).

Mizumoto, M., Pictorial representations of fuzzy connectives, Part II: cases of compensatory operators and self-dual operators. *Fuzzy Sets and Systems* **32(1)**, 45-79 (1989b).

Moore, J.T. and J.F. Bard, The mixed integer linear bilevel programming problem. *Operations Research* **38(5)**, 911-921 (1991).Murty, K.G., *Network Programming*. Prentice-Hall, Englewood, NJ (1992).

Negi, D., *Fuzzy Analysis and Optimization*. Ph.D. Dissertation, Dept. of Industrial Engineering, Kansas State University, Manhattan, Kansas (1989).

Negi, D. and E. S. Lee, Possibility programming by comparison of fuzzy numbers, *Computers & Mathematics Applications*, **25(9)**,43-50 (1993).

Nicholls, M.G., Aluminum porduction modeling - a nonlinear bilevel programming approach. *Operations Research* **43(2)**, 208-218 (1995).

Nicholls, M.G., The application of non-linear bi-level programming to the aluminum industry, *J. Global Optimization*, 8, 245-261 (1996).

ÓhÉígeartaigh, M., A fuzzy transportation algorithm. *Fuzzy Sets and Systems* **8(3)**, 235-243 (1982).

Önal, H., A modified simplex approach for solving bilevel linear programming problems. *European J. of Operational Research* **67(1)**, 126-135 (1993).

Önal, H., D.H. Darmawan and S.H. Johnson III, A multilevel analysis of agricultural credit distribution in East Java, Indonesia. *Computers and Operations Research* **22(2)**, 227-236 (1995).

Papavassilopoulos, G.P., Algorithms for static Stackelberg games with linear costs and polyhedra constraints. *Proceedings of the 21st IEEE Conference on Decision and Control*, 647-652 (1982).

Pearn, W.L., Solvable cases of the k-person Chinese postman problem. *Operations Research Letters* **16(4)**, 241-244 (1994).

Pedrycz, W., Fuzzy relational equations with generalized connectives and their applications. *Fuzzy Sets and Systems* **10**, 185-201 (1983).

Romero, C., A survey of generalized goal programming. *European J. of Operational Research* **25**, 183-191 (1986).

Saaty, T.L., *The Analytic Hierarchical Process*, 2nd ed., RWS Publications, Pittsburgh, PA (1990).

Sakawa, M., I. Nishizaki and Y. Uemura, Interactive fuzzy programming for multilevel linear programming problems. *Computers Math. Applic.* **36(2)**, 71-86 (1998).

Sakawa, M., I. Nishizaki and M. Hitaka, Interactive fuzzy programming for multi-level 0-1 programming problems through genetic algorithms. *European J. Operational Research.* **114**, 580-588 (1999).

Savard, G. and J. Gauvin, The steepest descent direction for the nonlinear bilevel programming problem. *Operations Research Letters* **15(5)**, 265-272 (1994).

Segall, R.S., An update on bi-level geometric programming: a new optimization method, *Applied Math. Modeling*, 17, 219-222 (1993).

Schrage, L., *LINDO: an optimization modeling system.* 4th ed., The Scientific Press, South San Francisco (1991).

Shafer, G., *A mathematical Reasoning of Evidence*, Princeton University Press, Princeton, NJ (1976)

Shi, X. and H. Xia, Interactive bilevel multi-objective decision making. *J. of Operational Research Society* **48**, 943-949 (1997).

Shih, H.S., Y.J. Lai and E.S. Lee, Fuzzy approach for multi-level mathematical programming problems. *Computers and Operations Research* **23(1)**, 73-91 (1996).

Shih, H.-S. and E.S. Lee, Fuzzy Multi-level Minimum-Cost Flow Problems. *Fuzzy Sets and Systems* **107(2)**, 159-176 (1999).

Shih, H.-S. and E.S. Lee, An Interactive Approach to Fuzzy Multi-Level Programming. *The Eighth International Fuzzy Systems Association World Congress (IFSA '99)*, 179-183, Taipei, Taiwan, ROC., August 17-20 (1999).

Shimizu, K. and E. Aiyoshi, A new computational method for Stackelberg and min-max problems by use of a penalty method, *IEEE Trans. Automatic Control*, AC-26, 460-466 (1981).

Shimizu, K. and Y. Ishizuka, Optimality conditions and algorithms for parameter design problems with two-level structure. *IEEE Transactions on Automatic Control* **AC-30(10)**, 986-993 (1985)

Shimizu, K, Y. Ishizuka, and J.F. Bard, *Nondifferentiable and Two-level Mathematical Programming.* Kluwer Academic Publishers, Boston (1997).

Shimizu, K. and M. Lu, A global optimization method for the Stackelberg problem with convex functions via problem transformations and concave programming. *IEEE Transactions on Systems, Man and Cybernetics* **SMC-25(12)**, 1635-1640 (1995).

Simaan, M. and J.B. Cruz Jr., On the Stackberg strategy in nonzero-sum games. *J. of Optimization and Applications* **11**, 544-555 (1973).

Simon, H.A., *Administrative Behavior.* 3rd ed., Free Press, New York (1976).

Soismaa, M., A note on efficient solutions for the linear bilevel programming problem, *European J. Operational Research*, 112, 427-431 (1999).

Sugeno, M. and T. Takagi, Multi-dimensional fuzzy reasoning, *Fuzzy Sets and Systems*, **9**, 313-325 (1983).

Taha, H.A., *Operations Research.* Prentice Hall, Englewood Cliffs, NJ, 5th ed. (1995).

Takagi, T and M. Sugeno, Fuzzy identification of systems and its application to modeling and control, *IEEE Trans. On Systems, Man, and Cybernetics*,**15(1)**,116-132 (1985).

Tanaka, H. and K. Asai, Fuzzy linear programming problems with fuzzy numbers. *Fuzzy Sets and Systems* **13(1)**, 1-10 (1984).

Tanaka, H., H. Ichihashi and K. Asai, Fuzzy decision in linear programming problems with trapezoid fuzzy parameters. in *Management Decision Support Systems Using Fuzzy Sets and Possibility Theory* (eds. by J. Kacprzyk and R.R. Yager), Verlag TÜV Rheinland, Köln, 146-154 (1985).

Thore, S., A. Nagurney and J. Pan, Generalized goal programming and variational inequalities. *Operations Research Letters* **12(4)**, 217-226 (1992).

Tiwari, R.N., S. Dharmar, and J.R. Rao, Priority structure in fuzzy goal programming. *Fuzzy Sets and Systems* **19**, 251-259 (1987).

Tobin, R.L. and T.L. Friesz, Formulation and solving the spatial price equilibrium problem with transshipment in terms of arc variables. *J. of Regional Science* **23(2)**, 187-198 (1983).

Tuy, H. A. Midalas and P. Varbrand, A quasiconcave minimization method for solving linear two-level programs, *J.Global Optimization*, 4, 243-263 (1994).

Tuy, H. A. Migdalas and P. Varbrand, A global optimization approach for the linear two-level program, *J. Global Optimization,* 3, 1-23 (1993).

Ünlü, G., A linear bilevel programming algorithm based on bicriteria programming. *Computers and Operations Research* **14**, 173-179 (1987).

Vicente, L.N. and P.H. Calamai, Bilevel and multilevel programming: a bibliography review. *J. of Global Optimization* **5(3)**, 291-306 (1994).

Vicente, L., G. Savard, and J. Júdice, Descent approaches for quadratic bilevel programming. *J of Optimizations Theory and Applications* **81(2)**, 379-399 (1994).

Vicente, L., G. Savard, and J. Júdice, Discrete linear bilevel programming problem. *J. of Optimization Theory and Applications* **89(3)**, 597-614 (1996).

von Stackelberg, H., *The Theory of the Market Economic*. Oxford University Press, London (1952).

Wang, Z., H. Nagasawa, and N. Nishiyama, A method for generating nondominated solution to a multiobjective-headquarters three-level decentralized system. in *Proceedings of 16th International Conference on Computer and Industrial Engineering,* Ashikaga Institute of Technology, Ashikaga, Japan, 536-539 (1994).

Wen, U.P., *Mathematical Methods for Multilevel Linear Programming*. Ph.D. thesis, State University of New York at Buffalo, NY (1981).

Wen, U.P. and W.F. Bialas, The hybrid algorithm for solving the three-level linear programming problem. *Computers and Operations Research* **13(4)**, 367-377 (1986).

Wen, U.P. and S.T. Hsu, A note on a linear bilevel programming algorithm based on bicriteria programming. *Computers and Operations Research* **16**, 79-83 (1989).

Wen, U.P. and S.T. Hsu, Linear bi-level programming problems - a review. *J. of Operational Research Society* **42**, 125-133 (1991).

Wen, U.P. and A.D. Huang, A simple tabu search method to solve the mixed-integer linear bilevel programming problem. *European J of Operational Research* **88(3)**, 563-571 (1996a).

Wen, U.P. and A.D. Huang, A tabu search approach for solving the linear bilevel programming problem. *J. of the Chinese Institute of Industrial Engineers* **13(2)**, 113-119 (1996b).

Wen, U.P. and S.F. Lin, Finding an efficient solution to linear bilevel programming problem: an effective approach. *J. of Global Optimization* **8**, 295-306 (1996).

Wen, U.P. and Y.H. Yang, Algorithms for solving the mixed integer two-level linear programming problem. *Computers and Operations Research* **17**, 133-142 (1990).

Werners, B.M., Aggregation models in mathematical programming. in *Mathematical Models for Decision Support* (ed. by G. Mitra), 295-305, Springer-Verlag: Berlin (1988).

White, D.J., A penalty function approach for solving bi-level linear programs, *J. Global Optimization*, 3, 397-419 (1993).

White, D.J., Solving bi-level linear programmes, *J. Math. Analysis Applications*, 200, 254-258 (1996).

White, D.J., Multilevel programming, rational reaction sets, and efficient solutions. *J. of Optimization Theory and Applications* **87(3)**, 727-746 (1995).

White, D.J., Penalty function approach to linear trilevel programming. *J. of Optimization Theory and Applications* **93(1)**, 183-197 (1997).

White, D.J. and G. Anandalingam, A penalty function approach for solving bi-level linear programs. *J. of Global Optimization* **3(4)**, 394-419 (1993).

Whitford, D.T. and W.J. Davis, A generalized hierarchical model of resource allocation. *Omega* **11(3)**, 279-291 (1983).

Wolfe, P., Simplex method for quadratic programming, *Econometrica*, 27, 382-398 (1959).

Yager, R.R., Fuzzy decision making including unequal objectives. *Fuzzy Sets and Systems* **1**, 87-95 (1978).

Yager, R.R., On a general class of fuzzy connectives. *Fuzzy Sets and Systems* **4(3)**, 235-242 (1980).

Yager, R.R., On ordered weighted averaging aggregation operators in multicriteria decisionmaking. *IEEE Transactions on Systems, Man, and Cybernetics* **18(1)**, 183-190 (1988).

Yager,R.R, S. Ovchinnikov, R.M.Tong and H.T. Nguyen, *Fussy Sets and Applications: Selected Papers by L.A.Zadeh*, Wiley, New York (1987).

Yano, H. and M. Sakawa, A three-level optimization method for fuzzy large-scale multiobjective nonlinear programming problems, *Fuzzy Sets and Systems*, 81, 141-155 (1996).

Ye, J.J. and D.L. Zhu, Optimality conditions for bilevel programming problems. *Optimization* **33**, 9-27 (1995).

Ye, J.J. and D.L. Zhu, A note on optimality conditions for bilevel programming problems. *Optimization* **39**, 361-366 (1997).

Yu, P.L., A class of solutions for group decision problems. *Management Science* **19(8)**, 936-946 (1973).

Zadeh, L.A., Fuzzy sets. *Information and Control* **8**, 338-353 (1965).

Zadeh, L.A., Fuzzy sets as a basis for a theory of possibility. *Fuzzy Sets and Systems* **1(1)**, 3-28 (1978).

Zadeh, L.A., *Fuzzy Sets and Applications:Selected Papers by L.A.Zadeh*, R.R.Yager, S.Ovchinnikov, R.M,Tong, H.T,Nguyen (eds), Wiley, New York (1987).

Zeleny, M., *Multiple Criteria Decision Making*, McGraw-Hill, New York (1982).

Zhu, Q, *Evidential Reasoning in Logic Systems*, Ph. D. Dissertation, Kansas State University, Manhattan, Kansas (1992)

Zimmermann, H.-J., Description and optimization of a fuzzy system. *International J. of General System* **2**, 209-216 (1976).

Zimmermann, H.-J., Fuzzy programming and linear programming with several objective functions. *Fuzzy Sets and Systems* **1**, 45-55 (1978).

Zimmermann, H.-J., Applications of fuzzy set theory to mathematical programming. *Information Sciences* **36**, 29-58 (1985).

Zimmermann, H.-J., Interactive decision support for semi-structured mathematical programming problems. in *Mathematical Models for Decision Support* (ed. by G. Mitra), 307-319, Springer-Verlag: Berlin (1988).

Zimmermann, H.-J., *Fuzzy Set Theory and Its Applications*. Kluwer, Boston, 3rd edition (1996).

Zimmermann, H.-J. and P. Zysno, Latent connectives in human decision making. *Fuzzy Sets and Systems* **4(1)**, 37-51 (1980).
Zimmermann, H.-J. and P. Zysno, Decision and evaluations by hierarchical aggregation of information. *Fuzzy Sets and Systems* **10**, 243-260 (1983).
Zionts, S., Some thoughts on research in multiple criteria decision making. *Computers and Operations Research* **19(7)**, 567-570 (1992).

Index

MIX
Papier aus verantwortungsvollen Quellen
Paper from responsible sources
FSC® C105338

If you have any concerns about our products,
you can contact us on
ProductSafety@springernature.com

In case Publisher is established outside the EU,
the EU authorized representative is:
Springer Nature Customer Service Center GmbH
Europaplatz 3, 69115 Heidelberg, Germany

Printed by Libri Plureos GmbH
in Hamburg, Germany